Liaison Functions 2013

Core Programming Area at the 2013 AIChE Annual Meeting: Global Challenges for Engineering a Sustainable Future

San Francisco, California, USA
3 – 8 November 2013

ISBN: 978-1-63439-044-6

Printed from e-media with permission by:

Curran Associates, Inc.
57 Morehouse Lane
Red Hook, NY 12571

Some format issues inherent in the e-media version may also appear in this print version.

Printed by Curran Associates, Inc. (2014)

AIChE
3 Park Avenue
New York, NY 10016-5991

Phone: (203) 702-7660
Fax: (203) 775-5177

www.aiche.org

Additional copies of this publication are available from:

Curran Associates, Inc.
57 Morehouse Lane
Red Hook, NY 12571 USA
Phone: 845-758-0400
Fax: 845-758-2634
Email: curran@proceedings.com
Web: www.proceedings.com

TABLE OF CONTENTS

(19a) AIChE 101

Gawel, K. J., Mondelez International
Lewis, M., UOP
Originally presented on: 11/3/2013 15:30:00 - 15:55:00

This session will provide an overview of the operating structure of AIChE including an overview on
each of the operation councils. This session will also discuss divisions, committees, and local sections.
In addition to giving an overview of each of these groups, the session will also go into detail about the
different ways you can get involved in one or many of these groups. This session will be of particular
interest to Young Professionals and those who are new to AIChE membership or involvement in AIChE
on a national level.

(19b) Navigating the AIChE Annual Conference

Gawel, K. J., Mondelez International
Lewis, M., UOP

Originally presented on: 11/3/2013 15:55:00 - 16:45:00

This session will review the logistics of the conference including understanding the program book
and how sessions are organized. It will also highlight key networking opportunities throughout the
conference. In addition to reviewing the logistics of navigating the conference, it will go through a
brief overview of how the conference is put together and how you can become a speaker or session
chair. The second half of the session will be devoted to questions and answers as well as a networking
opportunity. This session will be of particular interest to Young Professionals and anyone who is
attending a National AIChE Conference for the first time.

(59a) Sustainability Metrics For Biofuel Production Process Design

Engelberth, A. S., University of Maine
Mosier, N., Purdue University
Originally presented on: 11/4/2013 8:30:00 - 8:55:00

Sustainability is an important concept for chemical engineering process design. In process design for biofuel production, this is even more critical as the Renewable Fuel Standard (RFS 2) classifies biofuels on the basis of meeting specific environmental sustainability standards. This paper will present an example of how process design software can be used as a tool to develop estimates for sustainability metrics. The metrics that will be used to compare a sustainable design include, but are not limited to: capital cost, operating cost, heat integration potential, and equipment material. Additionally, we will present how feedstock supply chain considerations can significantly affect the environmental sustainability biofuel production processes.

(59b) Novel Framework To Examine Sustainability Of Biorefining Processes

Sukumara, S., University of Kentucky
Seay, D. J., University of Kentucky
Originally presented on: 11/4/2013 8:55:00 - 9:20:00

As the recent research is exploring possibilities for developing renewable bio-based fuel, there arises a need for a supporting model that guides the existing research to the appropriate path of process and product design. Biomass as a potential feedstock raises concerns on long term feasibility of this nonconventional method amidst increasing greenhouse gas emissions and growing demand for energy and transportation fuel. Considering the challenges inherited due to the inclusion of biomass as feedstock, a model is developed, demonstrating an innovative multidimensional approach to create a framework for comparing various processes and analyzing their upstream and downstream supply chain effects. This concept shows developing, interfacing and linking of various multidisciplinary optimization models utilizing supply chain modeling and process systems engineering skills. To illustrate its applicability, a case study is performed in the Jackson Purchase region in Western Kentucky. Initially, the model developed will encompasses various financial parameters involved in major conversion techniques pursued in the given region of interest. On evaluating the locally available biomass feedstock, thermochemical and biochemical processes for the production of energy and transportation fuels weredetermined to be the possible options. Subsequently, this analysis involving variable costs and capital investments will be appraised for its environmental and societal impact, capturing all possible aspects of sustainability. The resulting model is unique and informative in assisting policy makers and various stakeholders for evaluating sustainability of biofuels in any given region. Finally, the model can be applied to many other regions due to its generalizability, making it a powerful tool for assessing multiple facets of sustainability.

(59c) Case Studies For Sustainable Process Development In Biotechnology and Engineering Industry

Sammons, N. E. Jr., Auburn University

Originally presented on: 11/4/2013 9:20:00 - 9:45:00

Sustainable process development is quickly becoming a cornerstone in light of decreasing profit margins in conventional chemical process industries. Starting from conceptual modeling, it is imperative to utilize process integration, economic analysis, and environmental impact assessment to ensure that newly developing industries will meet investor expectations while performing in safe and environmentally friendly ways. Several industrial case studies will be presented to highlight the use of sustainable process development and their impacts on economic and environmental performance. More effective catalyst carriers could be utilized to reduce the footprint of reaction systems that involve intense amounts of heat transfer, while air filtration design in combination with full cost accounting models have shown promise in the development of green buildings and systems. Fermentation and protein purification development have demonstrated vast potential in carbon dioxide capture and replacement of petroleum-based feedstocsk for production of chemicals and fuels. Utilization of crop residues as enhanced value-added streams result in increased profitability in growing biomass-based industries, while the consideration of environmental factors in food production have led to the emergence of new products and markets with less reliance on animal farming.

(71a) I'm a Chemical Engineer. Why Should I Become a PE?

Parrish, W. R., Consultant
Jensen, C. D., Colorado School of Mines
Originally presented on: 11/4/2013 8:30:00 - 8:55:00

Many chemical engineers think Professional Engineering licensure is needed only when they are required to stamp drawings. As a result relatively few practicing chemical engineers consider obtaining Professional Engineer licensure. This presentation briefly outlines the value of becoming licensed from both the personal and business perspective. It presents the most common areas of engineering practice where having a PE can be beneficial or even mandatory by state law.

Licensure is based upon education, examination and experience. The presentation discusses both current and possible education requirements. It discusses the variability of experience requirements between various states. The presentation gives an overview of the Fundamentals Exam (FE) and Principles and Practice of Engineering exam (PE) required for licensure. It briefly discusses the process for making sure the exams are relevant and graded fairly. The FE exam will be making major changes in 2014 and an overview of these changes will be presented.

(71b) Fundamentals of Engineering Examination and Enhancing Student Performance

Peters, R. W., University of Alabama at Birmingham
Kirby, J. T., University of Alabama at Birmingham
Hosch, I.,
Originally presented on: 11/4/2013 8:55:00 - 9:20:00

Graduating students, in the present global economic environment/job market, face increasing difficulty and scrutiny while trying to secure their first engineering positions in the job market. Among the many entry-level requirements, employers are increasing mindful of a student's licensure potential. While favorable for the majority of engineering disciplines, civil and environmental engineering graduates are all but required to secure this certification prior to job placement. In an effort to increase student performance on this crucial examination, University of Alabama at Birmingham (UAB) Civil Engineering developed a FE style review course for graduating seniors in 2009. The student performance on the FE test (before and after implementation of the FE style review course) are highlighted. This paper will explore the ramification of this continuing effort and offer comparisons of pre/post course treatment. Furthermore, lessons learned and a perspective on student motivations will be highlighted.

(71c) The New Computer-Based FE Exam for Chemical Engineers
Mcdowell, D.,
Originally presented on: 11/4/2013 9:20:00 - 9:45:00

Beginning January 1, 2014, NCEES will offer all fundamentals exams in a computer-based format exclusively through approved Pearson VUE testing centers. This transition has been in development for almost three years and has led to many significant changes to include the decision to develop 7 free-standing, discipline-specific FE exams based on curriculum-specific content, a shorter exam length, year-round testing at locations throughout the country and quicker results.

This presentation will provide information on the NCEES content review process that was used to create the new stand-alone computer-based FE Chemical engineering exam and include a detailed comparison of the old and new exam specifications

(71d) Licensure - An Academic Perspective

Wickramasinghe, S. R., University of Arkansas
Spicer, T. O., University of Arkansas
Originally presented on: 11/4/2013 9:45:00 - 10:10:00

Most Chemical Engineering faculty members are not licensed. Why? Is it important for faculty to be licensed? A few departments have in the past required faculty teaching engineering courses to be licensed. Is this beneficial? Many undergraduates take the Fundamentals of Engineering exam but do not take the Professional Engineering exam. Why? Is the FE Exam relevant for assessing ABET outcomes? Is licensure important for faculty who are conducting research? Should we encourage faculty to obtain their PE license? In this presentation these questions will be discussed.

(71e) Use of the Fundamentals of Engineering Exam As An Engineering Education Assessment Tool

Rockstraw, D. A., New Mexico State University

Originally presented on: 11/4/2013 10:10:00 - 10:35:00

The Chemical Engineering program at New Mexico State University began using the Fundamentals of Engineering Exam as an assessment tool in 2000. It was found over the ensuing years that students in the program were not finding much value in taking the exam. Consequently, the sample size was always too small to provide any useful assessment information. The NCEES report "Using the Fundamentals of Engineering (FE) Examination to Assess Academic Programs" by LeFevre, et al. suggested use of a "scaled score" method to treat the FE data for use as an assessment tool. Collected data was treated by this method, but again data provided little value as the error bars on the scaled scores were of extremely large size relative to the scale of the scores. It was noted by the Ch E faculty that analysis by this method assumes the results are for a population rather than a sample, and does not address whether the sample taking the exam reflect the population.

It is assumed that the FE exam represents a national norm capable of providing a useful assessment of the effectiveness of a program. It was thus suggested that the curriculum be modified so that taking the FE exam is a requirement of graduation. In this manner, the true population would be reflected by the data, and a method of treating the data to perform an assessment could be developed. Because students do not have to pass the exam as a degree program requirement, concern was expressed that students would not put forth a valid effort. A protocol was developed to moderate this concern. Beginning in 2007/08, students in the program were required to sit for the exam during the fall semester of the senior year, with the registration fee paid by the department. Students who do not pass the exam must take the FE a second time at their own expense in the spring semester. The requirement to take the exam a second time at their own expense will help to ensure students will put forth their best effort to pass. In this manner, a delay in graduation timeframe is avoided, yet the department collects more useful data.

Data has been collected by this method for a few years, and analysis of the results have permitted identification of topic-specific strengths and weaknesses in the program. Results and the ensuing student support network will be discussed.

(85a) An Example Of Success Achieved With The Collaboration Mexico - Canada: A Scientific Project Sponsored By Conacyt - Mexico and Its IMPACT On Students

Serrano, B., Universidad Autonoma de Zacatecas.
de Lasa, H. I., University of Western Ontario
Moreira, J., University of Western Ontario
Vazquez, J. L. Sr.,
Hernandez, M. A.,
Guevara, S.,
Garcia Bermudez, S.,
Originally presented on: 11/4/2013 8:30:00 - 8:50:00

In November 2012, the National Science and Technology Council of Mexico (CONACYT) selected as a case of success the completed project 83144, Basic Science Call, 2007 – 01. The project's title is "Photocatalytic Reactors with Different Scale and Novel Photocatalysts for the Water Decontamination", being Dr. Benito Serrano Rosales the technical responsible. In addition, this project was presented in a conference in Can Cun, Mexico, with title "National Conference of Research of Basic Science, Successful Cases, CONACYT, Calls 2006 and 2007".

The general objective of the proposal was to synthesize, characterize and test novel catalysts doped or impregnated with metals, in the photodegradation of organic pollutants in water by using solar and UV light. Reactors of different sizes, 7, 35 and 300 liters, are used to establish reaction mechanisms, reaction rates, irradiation fields, and reactor efficiencies.

This research project was regarded as a success case because six papers were published in high ranking journals (three in preparation), two books as invited editors with one of them being re edited in China) and one book chapter, and because the following reasons:

Technical reasons:

The following original objectives planned in the protocol were achieved and published in the papers, books and presented in conferences and preceding: 1) Synthesize, characterize and test several photo catalysts, with different model compounds. 2) Propose a Unified Kinetic Model to describe the photo degradation of phenol, using six photo catalysts, in a broad range of concentrations and perform kinetic modeling. 3) Simulate the irradiation field of the Photo CREC Reactor using Monte Carlo. 4) Calculate the reactor efficiencies during the experiment span based on the hydroxyl radicals and intermediates. 5) Explain the effect of iron and other metals impregnated on the catalyst surface, during the degradation of phenol. 6) Use several metals when doping the catalyst.8) Utilize different reactor sizes in all experiments (7, 35 and 300 Liters). These objectives were achieved and published in papers, book chapters and presented in various conferences.

Academic Reasons: There was an impressive formation of human resources, thesis, summer research stays of students both in Mexico and Canada.

The following thesis were completed: Three doctoral, two master in science and eight undergraduate and the following thesis are in preparation: Four doctoral, three master in science and seven undergraduate. The work was divulgated in 23 conferences, Mexican and international.

There was mobility of students, researchers, faculties. The students sent to perform research stays: 1) In CREC – UWO - Canada, two graduate, six undergraduate. 2) In Universite Laval, Quebec City, four graduate, one undergraduate. Also, several students from Queretaro, Cohauila, performed summer research stays in the laboratory of UAZ.

National and international relationships with other universities.

The relationships with Mexican, Canadian and American universities were reinforced and increased. The following universities and researchers were involved:

1.- Universite Laval, Quebec Canada, Dr. Serge Kaliaguine. Involved students: Sarai Guevara Castillo (UAZ), Alfonso Talavera López (UAM-I), Omar Valdez Martinez (UAZ), Juan Carlos Paniagua Rodriguez (UMSNH).

2.- The University of Western Ontario, London Ontario Canada, Dr. Hugo de Lasa. This university was the main collaborator, and seven Mexican students have performed summer research stays, three Mexican students completed doctoral studies, and three students are pursuing the doctorate. Due this academic and scientific relationship we could publish papers, books, obtain a grant from IDRC-Canada to launch the Institute of Advanced Studies Canada – Mexico (IASCM) and to organize conferences in Mexico and Canada.

3.- Universidad Autonoma Metropolitana – Iztapalapa, Dr. Gustavo Fuentes and Dra. Gretchen Lapidus, the following students were involved: Alfonso Talavera Lopez and Oscar Joaquin Solis Marcial.

4.- Universidad Michoacana de San Nicolas de Hidalgo, Dr. Rafael Maya Yescas, student: Juan Carlos Paniagua Rodriguez.

5.- Universidad Autonoma del Estado de Mexico, Dra Reyna Natividad and Dr. Armando Ramirez Serrano, students Eduardo Martin del Campo and Oscar Alvarado.

6.- Universidad Autonoma de San Luis Potosi, Dr. Edgar Moctezuma Velazquez, student Alfonso Pinedo Escobar.

7.- Universidad Nacional Autonoma de Mexico, Dr. Camilo Arancibia Bulnes and Dr. Rosa Ramirez Zamora, students Sayra Lissette Orozco Cerros, Sergio Cuevas and Mario Alberto Hernandez Mazatan.

8.- University of Tennesee Campus Chatanooga, Dr. Jim Henry, students Jose Alberto Gonzalez Guerrero, Adan de Jesus Zacarias Balderas.

It is possible to see that the technical aspects were fulfilled and exceeded; this was a proposal of basic science, with the main goals of generating scientific knowledge. Also, a very important point was the extension of the Heterogeneous Photocatalysis for water to the purification of air and hydrogen production through water dissociation. Based on the results and development of this project, these topics were proposed and initiated as new research lines, taking advantage of the experience and existing infrastructure. Also, several additional projects were written and approved due this project.

It is considered that this proposal had its most important impact in the development and formation of the involved students, inasmuch as undergraduate students were interested in scientific research and were motivated to pursue graduate studies. The graduated students were encouraged to continue with the doctoral program, they met professors and researchers from other universities, institutions, countries, and this eventually will serve to them to get jobs.

The success of this proposal was based on the enthusiastic participation of researchers, students and collaborators, but was facilitated by the excellent academic technical and scientific collaboration between UAZ and UWO – Canada, enhanced and promoted by the IASCM, created in 2005, an institute whose objectives are (among others): 1) Develop collaborative research in areas of strategic importance for both Canada and Mexico. This would consist of developing, testing and implementing technologies for the protection of the environment; the use of clean and renewable energy resources; as well as research on related socio-cultural, health, legal and migration issues. 2) Contribute to the training of Canadian and Mexican Master and PhD graduate researchers and university professors, in their respective fields.

This success case is an example of the importance of the international collaboration and its multiplicative effect on the formation of human resources, which leads to obtain more productivity in technical, scientific and academic areas.

(85b) Dynamic Design of a Cryogenic Air Separation Unit

Schmidt, S.,
Clayton, R.,
Originally presented on: 11/4/2013 8:50:00 - 9:10:00

Abstract

We present the dynamic design of an air separation unit (ASU), which is used to separate oxygen, nitrogen, and argon products from air at very low temperatures. This unit can be used as an isolated plant to produce the products to be sold or as a unit joined to a larger plant, also called a "piggy back" plant, to produce products to be used in an additional process [1]. These gases are commonly used in refineries and oil recovery efforts. This process is convenient because the raw material is available at no cost and in limitless amounts. The primary cost in the process is compression. In order to minimize costs, heat integration is used throughout the process. This particular ASU will produce 1500 metric tons of 99.5% oxygen, 5000 metric tons of 99.5% nitrogen, and 58 metric tons of 95% crude argon each day to a customer. Heat integration will eliminate all costs from heating and cooling in the process. The compressor capital cost will be $16.5 MM[1]. The venture guidance appraisal for this process is $118.5 MM. When running at full capacity, the plant will sell $113.9 MM worth of products. With a total annual cost for equipment and utilities of $39.0 MM, this plant will yield a yearly profit of $73.4MM. An integrated control structure has been designed to maintain product purity and reject atmospheric (temperature and pressure) disturbances, as well as flowrate disturbances. An overview of both a feed control and on demand scheme is included in the following pages, along with disturbance testing for both schemes.

(85c) Navigating the Natural Gas Boom: Policy Option for Ensuring a Safe, Clean, and Economical Natural Gas Revolution

Crothers, A., RWTH Aachen University
Winkler, P., Air Products and Chemicals, Inc. (Retired)
Originally presented on: 11/4/2013 9:10:00 - 9:30:00

The recent revolution in natural gas has revealed incredible economic opportunity for this nation. There are copious natural gas reserves in the US, particularly in shale repositories. Capturing these reserves will yield massive economic growth in direct energy production, transportation, and downstream product synthesis industries. Despite these benefits, there are continual controversies on how best to utilize these newfound resources and how to mitigate the potentially severe environmental implications. It is important that policy makers decided on how best to incentivize safe shale gas mining, regulate global trade, and promote responsible utilization of these limited resources. Obtaining natural gas from shale repositories requires hydraulic fracturing which has controversial public health and environmental implications. Moreover, while gas is cleaner than coal and oil, it does not provide a completely nonpolluting and sustainable solution our scarce energy supplies problem. Providing regulatory certainty to these challenges of natural gas would provide a strong foundation for growth.

(85d) A Wavelength Shifting Polymer to Enhance the Growth and Production of Cyanobacteria

Nelson, T.,
Salem, D. R., South Dakota School of Mines and Technology
Winter, R. M., South Dakota School of Mines and Technology
Originally presented on: 11/4/2013 9:30:00 - 9:50:00

Cyanobacteria consume carbon dioxide while producing oxygen and desired genetically engineered products. When grown in photobioreactors, cyanobacteria can provide both oxygen and fuel during long term space travel and extraterrestrial colonization. The overall goal of the research team is to optimize the production of linalool, a terpene alcohol, by the cyanobacteria genus anabaena grown in a photobioreactor. The goal of this particular sector of the research is to modify the light delivered to the anabaena in order to stimulate growth rate and production of fuel. Polymer films, poly(methyl methacrylate), were prepared with fluorescent molecules which absorb photons of non-essential wavelengths and shift them to wavelengths more beneficial to cyanobacteria photosynthesis. These films were prepared by using a solvent casting method and tested for fluorescence emissions. Ultraviolet light was converted by Coumarin 460 to match the 440 nm absorption peak of anabaena (chlorophyll a). Green light (approx. 532 nm) was red-shifted to higher wavelengths by inexpensive Rhodamine fluorescent dyes. These energy shifts are expected to increase the cyanobacteria growth and alcohol production rate by increasing the available specific energy required by these processes. A current study is being done to observe growth rate, by optical density, and fuel production, by HPLC, of the anabaena when stimulated with particular LED lighting arrays characterized by a spectroradiometer. Latter tests will use a solar spectrum of light that is modified by the wavelength shifting polymer. Results of this study will indicate the quintessential wavelengths for maximum fuel production. Future work includes up-conversion of infrared light using triplet-triplet annihilation and increasing fluorescent efficiencies by surface plasmon resonance from metallic nanoparticles and films. This wavelength shifting technology could be used in other applications that enhanced photosynthesis would benefit, such as other photo-reactors, crop production, and development of LEDs used for grow lights.

(85e) Exploring Multiphase Viscous Drop Impact With a Bulk Fluid

Logia, A. N., Stanford University
Walker, T. W., South Dakota School of Mines and Technology
Fuller, G. G., Stanford University
Originally presented on: 11/4/2013 9:50:00 - 10:10:00

Exploring Multiphase Viscous Drop Impact with a Bulk Fluid

Miscible fluid systems are all around us from the mixture of water and dish soap in the kitchen to the dissolution of oxygen into blood. Building our intuition about how miscible and immiscible fluid systems differ can significantly further our understanding of fluid mechanics.

We have developed an experiment to explore how a droplet impacts a bulk fluid interface. A drop of dense, viscous fluid is ejected from a syringe, free falling into a bath of water. A high-speed video is captured by triggering a laser photo diode system. These liquid-liquid drop impact systems can be inherently complex due to the time dependent interfacial tension between the two phases. Their interactions can produce a variety of unique shapes and morphologies including toroids, liquid columns, and captured bubbles. We aim to understand the conditions under which these various regimes form and the physics governing their formation.

One particularly interesting regime is the "free-surface pendant drop." Under certain conditions, drops will partially wet the bath surface, residing on top of the bath despite the drops greater density. Interfacial tension between the drop and the bath reduces, as the drop dissolves into the bath, and the drop is driven down into the bath once gravity overcomes the interfacial tension. Between the surface of the bath and the bulk of the downward driven drop, a liquid column is formed. The liquid columns formed by these "free-surface pendant drops," as well as many other regimes, could have a variety of practical applications in industry.

(85f) Designing Copper Zinc Oxide as a P-Type Transparent Conducting Oxide for Photovoltaic Applications

Sun, D.,

Originally presented on: 11/4/2013 10:10:00 - 10:30:00

Transparent conducting oxides (TCOs) are electrically conductive materials that are used in thin-film opto-electronic applications, such as solar cells, but major improvements in performance and cost are necessary for TCOs to be used on a commercial scale. We present Copper-doped Zinc Oxide as a low-cost TCO candidate. We determined how the electrical transport properties and optical properties vary with composition by performing density functional theory calculations, including gradient-corrected, hybrid, and Green's functions. With increasing Cu concentration (x), the band gap of Copper-doped Zinc Oxide ($Zn_{1-x}Cu_xO$) decreases due to the down shift of the valence band. However, the optimal value of x is determined by the balance between the increase in electrical conductivity and the decrease in optical absorption. We also compare our results to ternary oxides formed from zinc and copper, and discuss strategies for designing materials with desirable properties.

(104a) Use Vesicle Membrane As a Designable Platform for Molecular Recognition ~ Toward Creation of Bio-Inspired Chemical Engineering Based On Membranome ~

Umakoshi, H., Osaka University
Suga, K., Osaka University
Originally presented on: 11/4/2013 12:30:00 - 13:00:00

A "*Biomembrane*" is a highly-organized self-assembly of biomolecules (i.e. lipid, protein etc.) and a key interface for the survival of biological cell. The "*Membranome*" can be defined as the properties of vesicle (or liposome), which arise from the bilayer molecular assembly of amphiphiles, focusing on "*emergent properties*" which are not present in the individual components, and is gradually recognized as an important research methodology to investigate the potential functions of vesicles (or liposome) and to apply them for the bioprocess design. A liposome possesses several benefits in the recognition of (bio)molecules, where it can recognize them with (i) electrostatic, (ii) hydrophobic interaction, and (iii) stabilization effect of hydrogen bonds in the hydrophobic lipid environment. A molecular recognition with high selectivity can be achieved by the simple liposome membrane on its surface through the combination of (i)-(iii). We are now focusing on the creation of "*Bio-Inspired Chemical Engineering*" based on the "*Membranome*" by selecting "liposome or vesicle membrane" as a "*platform*" to perform the "*recognition*" and "*conversion*" of biomolecules, such as (a) *Artificial Chaperone* , (b) A*rtificial Membrane Protein* , (c) *Artificial Enzyme*, (d) *Artificial Virus,* (e) *Artificial Organ* (liposome-immobilized hollow fiber membrane or liposome-immobilized hydrogel matrix) and etc. The author will make a presentation on the basic concept of "*membranome*", together with some examples.

(104b) Morphological Transitions in Block Copolymers Directed By Superparamagnetic Nanoparticles

Raman, V., Massachusetts Institute of Technology
Olsen, B. D., Massachusetts Institute of Technology
Hatton, T. A., Massachusetts Institute of Technology
Originally presented on: 11/4/2013 13:00:00 - 13:20:00

This computational and theoretical study investigates the self-assembly of superparamagnetic nanoparticles and block copolymers under external magnetic fields. A variety of morphological transitions are observed based on the field orientation, nanoparticle loading, and selectivity of the nanoparticles for the blocks. For symmetric block copolymers, chaining of superparamagnetic nanoparticles under in-plane magnetic fields is shown to achieve long range orientational order of the block copolymer nanodomains and is found to be dependent on nanoparticle size, volume fraction and magnetization strength. A critical selectivity of the particles for one nanodomain is observed, above which strong alignment results and below which comparatively disordered structures are formed. Higher magnetization strengths are found to reduce equilibrium defect densities in the nematic-isotropic ordering of lamellar thin films, as corroborated by scaling arguments. For asymmetric coil fractions forming hexagonal block copolymer nanostructure, the in-plane field induced chaining of the nanoparticles selective for the minority block, leads to the formation of stripe phases oriented parallel to the magnetic field. Furthermore, in-plane field induced chaining of nanoparticles selective for the majority block leads to alignment of hexagonal morphology with <100> direction oriented parallel to the external magnetic field. Out of plane magnetic fields induce repulsive dipolar interactions between the nanoparticles that annihilate the defects in the hexagonal morphology of the block copolymer when the nanoparticle is selective for the minority block. Honeycomb lattices are obtained using nanoparticles selective for majority block under out of plane magnetic fields for certain specific nanoparticle loadings. Commensurability of nanoparticle size and loadings with the block copolymer structure is critical in optimizing the ordering of the final composite.

(104c) Design of Autonomously Moving Colloidal Objects - Systematic Motion of Vesicles, Particles and Oil/Water Interface in a Homogenous Chemical Field -

Shioi, A., Doshisha University

Originally presented on: 11/4/2013 13:20:00 - 13:50:00

The second law of thermodynamics states that all systematic motions are transformed into a random motion with molecular length scale. In biological system, however, spontaneous motions driven by chemical reaction with a molecular length scale generates a systematic motion with macroscopic scale. To study how and why the systematic motion can evolve over the hierarchy from an apparent random dynamics with smaller length scale is fascinating for designing a biomimetic chemical system. Design of autonomously moving colloidal objects is one of the methodologies that enable to understand the mechanism for the energy transduction. Moreover, the study of self-moving system may develop into various applications related to biomimetics and microfluidics. Autonomously moving colloids can demonstrate numerous significant characteristics such as transduction of chemical potential into work without heat, chemosensitive motion, and pulse-like motion with periodicities responding to the chemical environment. Here, three types of colloidal systems that exhibit autonomous motions will be discussed. Chemosensitive motion of oil/water interface that is caused by chemical Marangoni instability is the first example. The convection caused by the instability can be rectified to make one dimensional motion of a float at the interface. Second, a vesicle with rhythmic shape change is discussed. This vesicle shows a sustainable motion under a pH gradient. A vesicle that swims in a solution with specific chemical is also shown. Finally, autonomous motion of catalytic particles is discussed. The motion of homogeneous catalytic particle can be controlled by its geometry.

(104d) Development and Properties of Novel Photo-Cleavable Surfactants

Sakai, H., Tokyo University of Science
Aikawa, S., Tokyo University of Science
Fukuda, K., Tokyo University of Science
Shrestha, R.,
Endo, T., Tokyo University of Science
Sakai, K., Tokyo University of Science
Abe, M., Tokyo University of Science
Originally presented on: 11/4/2013 13:50:00 - 14:20:00

Recently, we have reported a new cinnamic acid-type photo-cleavable surfactant, C4-C-N-PEG9 that experiences a photo-cleavage through UV-induced cyclization in aqueous solution, yielding a coumarin derivative and an aminated polyoxyethylene compound[1)]. Here, we have studied the effects of C4-C-N-PEG9 on the photorheological behavior of viscoelastic wormlike micelles formed by aqueous mixture of nonionic surfactants, polyoxyethylene phytosterol ether ($PhyEO_{20}$) and tetraoxyethylene dodecyl ether ($C_{12}EO_4$). When C4-C-N-PEG9 is added to the viscous, non-Newtonian 4.9 wt% $PhyEO_{20}/H_2O$ + 2.4 wt% $C_{12}EO_4$ fluid, a decrease in viscosity is observed, while upon UV irradiation C4-C-N-PEG9 added solution regains the viscosity. It increases by 200 times, suggesting a modulation in micellar shape and size. The UV-Vis spectroscopy and HPLC measurements confirm that the *trans*-C4-C-N-PEG9 with large head group and its photodecomposition to form less surface-active coumarin derivative and an aminated polyoxyethylene compound are responsible for different rheological behavior[2)].

References 1) H. Sakai, et al., *Journal of Colloid and Interface Science,* 376, 160-164, (2012).

2) H. Sakai, et al., *Langmuir,* Article ASAP, DOI: 10.1021/la4001939.

(104e) Ordering Transitions Triggered By Specific Binding of Vesicles to Protein-Decorated Interfaces of Thermotropic Liquid Crystals

Abbott, N. L., University of Wisconsin-Madison
Tan, L. N., University of Wisconsin-Madison

Originally presented on: 11/4/2013 14:20:00 - 14:40:00

This presentation will address an investigation of the interactions of phospholipid vesicles with protein-decorated interfaces of thermotropic liquid crystals (LCs). We have observed that following the capture of vesicles at a protein-decorated LC interface via specific binding interactions, phospholipids are transferred from the vesicles onto the LC interface to form a monolayer, partially displacing proteins from the LC interface and thereby triggering a continuous ordering transition in the LC. The dynamics of this process are accelerated substantially by the specific binding event relative to a protein-decorated interface of a LC that does not bind the ligands presented by vesicles. The observation of the *continuous* change in the ordering of the LC is significant as it is consistent with the presence of sub-optical domains of proteins and phospholipids on the LC interface. An additional significant hypothesis that emerges from the work to be reported in this presentation is that the ordering transition of the LC is strongly impacted by the bound state of the protein adsorbed on the LC interface, as evidenced by the influence on the LC of (i) "crowding" of the protein within a monolayer formed at the LC interface and (ii) aging of the proteins on the LC interface. Overall, these results demonstrate that ordering transitions in LCs can be used to provide fundamental insights into the competitive adsorption of proteins and lipids at oil-water interfaces, and that LC ordering transitions have the potential to be useful for reporting specific binding events involving vesicles and proteins.

(104f) Directed Aggregation and Fusion of Vesicles Induced By a DNA-Surfactant

Goto, M., Kyushu University

Originally presented on: 11/4/2013 14:40:00 - 15:00:00

Conventional synthetic strategies have provided a number of attractive molecular devices, but the sizes of the devices produced using these strategies are limited to a few hundred nanometers at most. Molecular self-assembly is an alternative approach for constructing systems that are larger (submicrometers to micrometers) and more complex. In particular, lipid vesicles, which are spherical bilayer shells encapsulating an aqueous interior, have been widely studied as a biomimetic example of molecular self-assemblies. Such lipid vesicles can range in size from 50nm to ~1mm, and contain various compounds, including metal ions, macromolecules and biomolecules. Furthermore, modification of the lipid bilayer can confer diverse functions onto lipid vesicles. Higher order or secondary levels of self-assembly of lipid vesicles have been achieved through vesicle aggregation using electrostatic and ligand–receptor interactions. These higher ordered assemblies of lipid vesicles were also studied as a simple model for biological tissues.

Over the last decade, oligonucleotides have attracted much attention as intelligent materials due to their highly specific and designable binding properties toward their complementary sequences. Many researchers have employed oligonucleotides as building blocks to construct nanostructured systems with well-designed geometric shapes and sizes. Besides, oligonucleotides have been used as molecular glues for cross-linking, immobilizing and aggregating proteins, nanoparticles, colloids and cells. Oligonucleotides were also introduced to the vesicles and utilized for the immobilization of vesicles on an electrode or a plate. Patolsky et al. utilized the DNA-tethered vesicles for amplification of hybridization signal. Yoshina-Ishii and Boxer first reported the sequence-specific immobilization of lipid vesicles. These studies suggested the utility of DNA-tethered vesicles. Most of them mainly focused on surface modification and nanofabricated structures, rather than on the construction of macroscopic structured systems. Wu et al. reported that an extremely high concentration (10mg/ml) of long linear DNA-induced non-selective aggregation of a small portion of zwitter ionic vesicles, where the aggregation was supposed to be induced not by the DNA hybridization but simply by the electrostatic interaction. Graneli et al. first reported the interaction among DNA-tethered vesicles on a DNA-modified gold substrate and succeeded in the formation of three-dimensional vesicle matrices based on DNA hybridization. There are, however, considerable uncertainties on the DNA-directed aggregation of vesicles in solution. For example, are the DNA-tethered vesicles responsive to external stimuli such as temperature, ionic strength, etc.? Here, we investigate DNA directed aggregation of vesicles in solution using DNA-surfactants and clarify if the DNA-tethered vesicle possesses the DNA-derived characteristics to be responsive to external stimuli. Furthermore, we successfully accomplished DNA-directed fusion of vesicles based on this assembly procedure, leading to the production of a micro-reactor involving an enzyme-catalyzed reaction in the giant vesicle produced.

(150a) Race To 2015: The Completion Of The Millennium Development Water, Sanitation, and Hygiene Goal

DeLago, C., SIT, AIChe (American Institute of Chemical Engineers) Chapter
Originally presented on: 11/4/2013 12:30:00 - 12:50:00

Today, one in every five people do not have access to clean water, especially those in third-world countries. Many rivers and lakes have now been contaminated by effluent factory waste causing them to have poisonous amounts of ammonium, phosphate, and arsenic. This, in addition to natural causes such as droughts, over population, and natural disasters have turned the need for water into a fast-growing problem that could soon turn into a global epidemic if proper precautions are not taken. With the turn of the new millennia in 2000, many of the world's leading countries knew something had to be done to aid developing nations through these water crises. As a result, in 2005, the United Nations put into play the Millenium Development Goals, with the seventh goal being to reduce the amount of people without access to clean water and proper sanitation by 50% before 2015. A portion of this goal was prematurely achieved in 2010 with 10 million people worldwide gained access to a sustainable water source.

I would like my project this summer to focus on this initiative by focusing on what the American government agencies and non-profit organizations have done thus far to aid the United Nations in completing this goal and conduct further research into what still needs to be implemented to fulfill the objective. Similarly, I would like to look into what other countries and organizations around the world have done to help meet this goal and conclusively postulate as to whether or not the Initiative will be completed in time. Lastly, I would like to focus on what the United States, the rest of the world, and The U.N. should do in the future to make sure that the Water and Santitaion Crisis remains an issue of the upmost importance.

(150b) Dynamic Design of a Ethylamine Production Plant

Becht, K.,

Wiedorn, A.,

Originally presented on: 11/4/2013 12:50:00 - 13:10:00

The ethylamine production facility documented in this process design produces on average 98 tons/day of triethylamine and 186 tons/day of diethylamine. Fresh ammonia and fresh ethanol are fed into the process with the presence of hydrogen to produce mono, di-, and tri- ethylamine with a byproduct of water. The products are produced in a series of three reactions. Ammonia and ethanol react to produce mono-ethylamine. Diethylamine is produced from monoethylamine and ethanol and triethylamine by diethylamine and ethanol. The unreacted ethanol and ammonia, the monoethylamine and a portion of diethylamine and triethalmine are recycled back into the reactor. This production facility produces a return on investment (ROI) of 42% and will make a profit after four years. The steady state plant design was created from optimizing cost and efficiency and consists of a compressor train, cooled plug flow reactor and a series of six distillation columns. The dynamic simulation of the process focuses on maintaining the product purities and controlling the temperatures due to safety concerns at high process temperatures. A HAZOP review was performed on the reactor temperatures since this is a critical safety concern. The safety of the plant and its operators is also discussed. Issues of flammability, auto ignition temperatures, and construction materials are addressed.

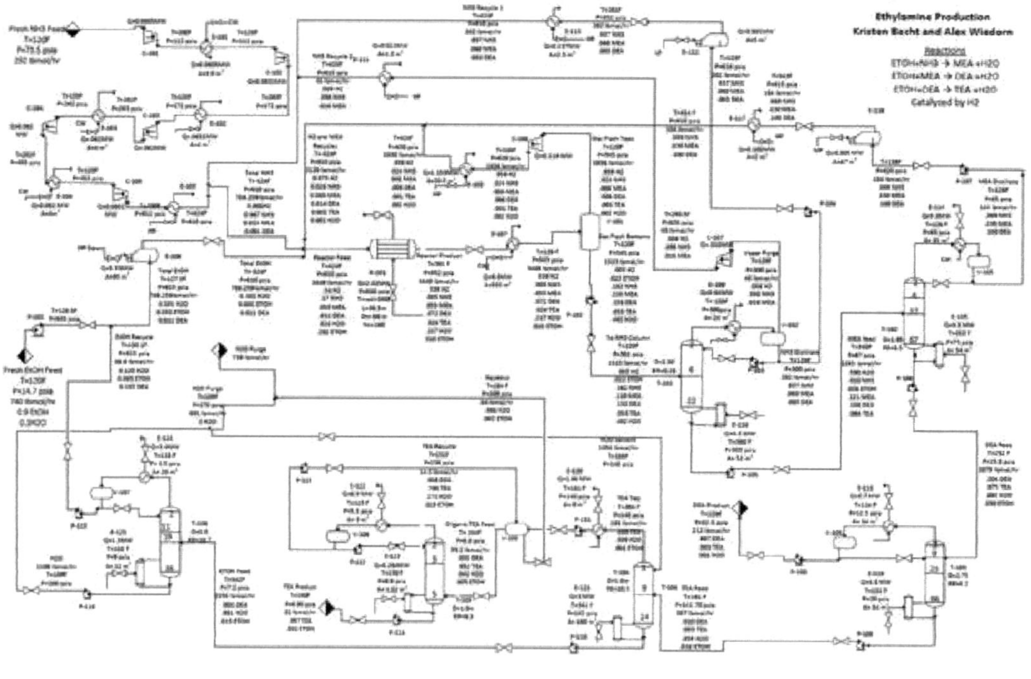

Figure 1. Optimized steady state design of the ethylamine production plant.

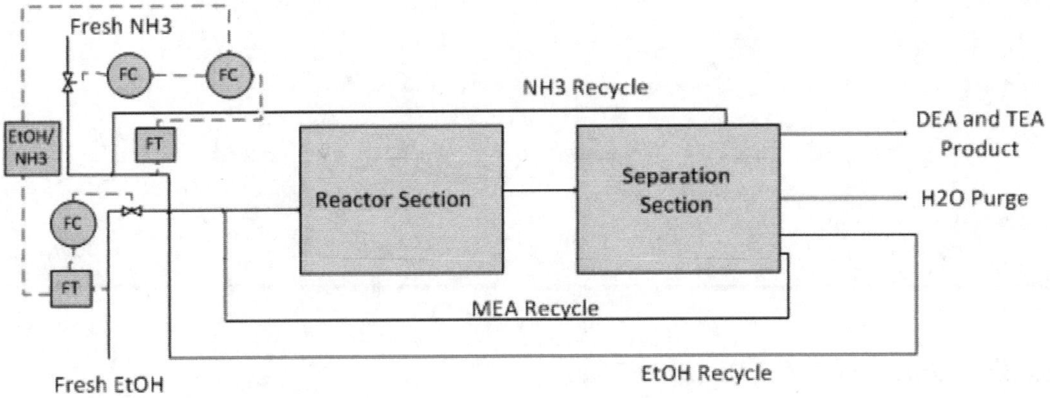

Figure 2. Control scheme for plant wide control (PWC).

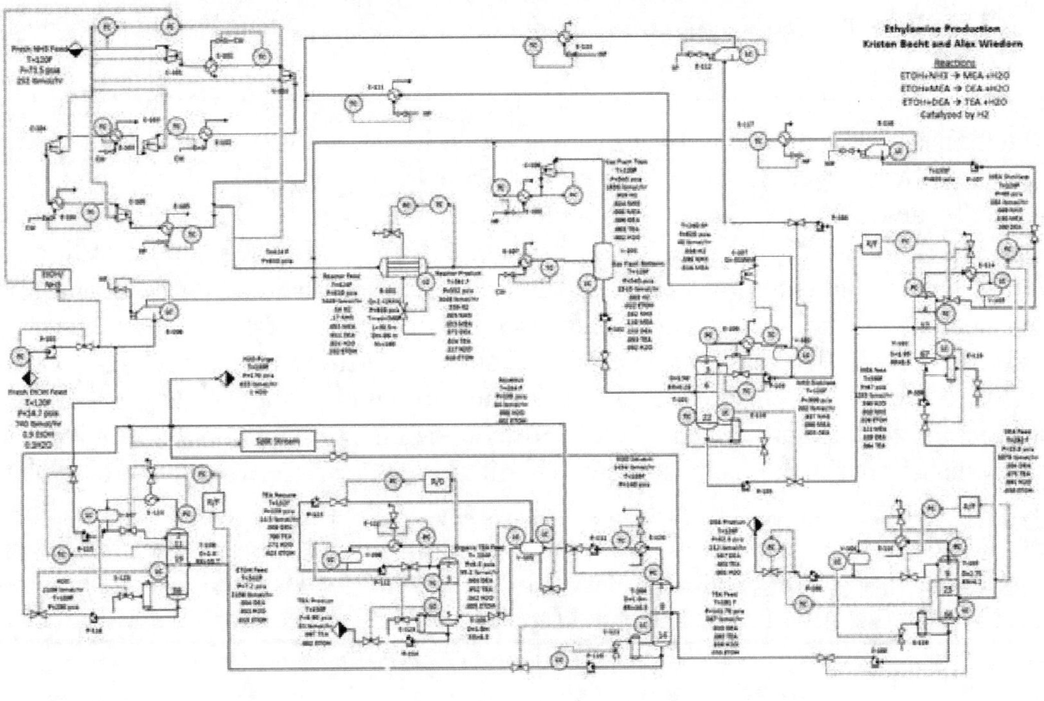

Figure 3. Dynamic design of the ethylamine production plant.

(150c) Acetic Acid and Methanol Recovery Unit

Berard, V. F., Lehigh University
DiNardo, A., Lehigh University
Fortney, S., Lehigh University

Originally presented on: 11/4/2013 13:10:00 - 13:30:00

The goal of this unit is to utilize a stream from the saponification section of a Polyvinyl Alcohol plant situated on the Gulf Coast of the United States. In addition to methanol to be recovered, the feed stream contains large quantities of methyl acetate. Therefore, methyl acetate is hydrolyzed to produce additional methanol and acetic acid for sale. The total amount of methanol is purified to 99.9 weight percent and acetic acid purified to 99 weight percent to meet prescribed specifications. Biodegradable sodium acetate also enters in the feed stream and is removed through a purge stream to be dumped. A second impurity, acetaldehyde, is removed from the plant and incinerated to eliminate the carcinogenic hazard. The optimized steady-state design includes five distillation columns and one reactor and three recycle streams. Unconverted methyl acetate is returned to the front-end with only trace amounts exiting the unit unconverted. An azeotrope between methyl acetate and methanol is broken using extractive distillation techniques, resulting in a second recycle stream for the water solvent. The final recycle arises from the excess water fed to the reactor to increase methyl acetate conversion. At full capacity, the plant recovers 177 million pounds of methanol per year and produces 101 million pounds of acetic acid for sale. Further, a plant wide control scheme is implemented to operate the plant. The control structure is designed to withstand composition disturbances of +/- 10%, a 10 % increase in feed rate, and a 33% decrease in feed rate. An economic analysis shows that the plant will begin to make a profit and consistently make a profit of at least $50 million for the following 10 years with a return on investment of 233%.

(150d) A Re-Examination of US Water Policies That Prevent Sustainable Water Consumption and Reuse

Kusanto, N., Oregon State University
Originally presented on: 11/4/2013 13:30:00 - 13:50:00

Water management and allocation has been integral in the success and development of the United States of America. While ground water has been historically recognized as a personal property right, surface waters required greater regulation. Thus, the first water policies were developed and reflected man's desire to harness the wild rivers for human development. Acts such as the 1920 Federal Power Act and 1922 Colorado River Compact illustrate governmental hopes of controlling water ways for power generation and agricultural allocation among states. The Colorado River Compact, in particular, attempted to fairly divide a shared water resource among users across a number of states based on projected need. As US expansion and development continued, the federal government sought to further centralize control and regulation of waterways through the 1936 Flood Control Act and the 1948 Water Pollution Act. Early water policy failed to protect water supplies, allowing US development to severely deplete and pollute water resources. As such, in the last quarter of the 20th century, focus was placed on protecting drinking water, 1974 Safe Drinking Water Act, and to support research which investigates how the country can safely reclaim and reuse waste water, Title XVI in 1992. Looking forward into the 21^{th}century, water policies should reflect a need for sustainable water consumption and increase of water reuse and reclamation, employing the discoveries made in science.

The present need and desire to encourage wise water use and water recycle must be driven by an understanding of the factors which have historically prevented municipalities and industry from water recycling. The current system for water distribution taxes users a nominal fee to subsidize the cost of water transportation. The water, itself, is free. As such, the abundance of free water coupled with lack of public knowledge on water management issues have contributed to excessive consumption of fresh water which has depleted water resources around the US. For example, the average depth in the 174,000 square mile Ogallala aquifer, one of the world's largest, has dropped from 240ft to approximately 80ft. Conservative estimates predict that full depletion of the aquifer will occur in the next 25 years, requiring a recharge time of 6,000 years to reach initial levels. Clearly, current time horizons dictating water use disregard long term water conservation. If the rate of water withdrawal continues to exceed the existing rate of aquifer recharge, the US will soon run out of a necessary resource which is at the heart of the US agricultural belt. Because the irrigated plains produce 20% of American grain and corn, failure to control water use will result in a collapse of the international agricultural market. Concerns for the protection of public health and a decline in public trust in government agencies are at the forefront of a list of challenges which prevent local governments from enacting infrastructure changes to improve water recycle. On a grander scale, the federal government is only able to act in a manner consistent with the U.S. Constitution. Thus, state laws may vary and can come into conflict if aquifers or surface water resources are shared. Recognizing how to address conflicts between states, municipalities, and the federal government is important to finding a solution. Industry must follow the regulations imposed by the three actors above. Strict water reuse regulations lead to cost-benefit analyses which prevent companies from conserving water and choosing to discharge their waste. Likewise, incentives are not currently in place to reward companies who are able to reduce consumption within limited time frames.

A simple mass balance on water usage in the US will illustrate that action must be taken to help conserve water resources. The national government should invest more heavily on alternative future analyses to determine the effects water use has on biophysical and human systems. Increasing scientific literacy in the public sector will also play a key role in gathering support from municipalities. Moreover, incentives, much like solar energy tax breaks, must be implemented to encourage better use of water. Infrastructure costs for water must be externalized to consumers resulting in increasing rates for water use, which will encourage sustainable water consumption. It is suggested that excessive water use be penalized on a tiered system to penalize waste and encourage sustainability. Existing policies must also be reexamined and redefined to reflect research discoveries. The relaxation of water reuse regulation coupled with more strict regulations on water discharge may encourage companies to conserve water. This paper will provide possible solutions for the aforementioned difficulties regarding water reuse.

(150e) Toluenediamine Production Through Hydrogenation in Methanol Solvent

Masel, R.,
Smith, D., Lehigh University
Originally presented on: 11/4/2013 13:50:00 - 14:10:00

A feasibility analysis for a new toluenediamine (TDA) production plant was conducted. The production process selected for investigation was the hydrogenation of dinitrotoluene (DNT) in liquid methanol solvent over a nickel catalyst. The product TDA and reactant DNT have much higher boiling temperatures than the solvent methanol and byproduct water, and the boiling points of the reactants and products are too close to allow for sufficient separation. Therefore, the reactor must be sized sufficiently to achieve nearly complete conversion, as the reactant cannot be separated and recycled. Oligomerization of the DNT and TDA occurs at higher temperatures, generating high boiling tars that deactivate the catalyst and contaminate the product streams. Additionally, the thermal instability of DNT, as well as TDA, tars, and other nitrogen containing byproducts necessitate a temperature limit of approximately 350 F in any streams that could potentially contain the least stable component DNT. The composition of water in the product must be no greater than 0.1 weight percent. Two potential alternatives in distillation of the solvent and water from the products were apparent as a result of both a temperature and composition specification in the bottoms product. The first was a single column at high vacuum requiring refrigeration in the condenser, and the second was two separators with a column at atmospheric pressure using cooling water in the condenser before the vacuum column achieving the desired composition. The total annualized cost (TAC) for the single column was $\$6.361/10^6$, while the TAC for the double column system was less at $\$4.696/10^6$ due to reduced refrigeration costs. Health and safety hazards were considered in a process safety analysis that outlined concerns for implementing the process. A product price of $1.00 per pound of TDA was calculated to yield sufficient return on investment (ROI) of approximately 51.9% with a payback period of 1.9 years. Production of toluene diisocyanate (TDI) from toluenediamine was considered as the primary use, with an estimated price in 2016 of $1.459 per pound of TDI, validating the assigned price of intermediate TDA. The net present value of the project considering five years of operation after startup is $56.2 MM, or $129.3 MM over an operation lifetime of 14 years.

(150f) Experimentation: The Foundation of Career Growth

Worstell, J. H., Worstell and Worstell, Consultants
Orsak, J.,
Arredondo, J., University of Houston

Originally presented on: 11/4/2013 14:10:00 - 14:30:00

Every chemical engineer will design experiments and evaluate experimental data sometime during their career. Thus, it is important for chemical engineering students to have a meaningful laboratory experience early in their scholastic career. At the University of Houston, chemical engineering students begin evaluating experimental data during their Junior year. This situation raises challenges for the laboratory staff. First, designing a laboratory course is difficult when you consider how little time Junior chemical engineering students have available for a laboratory course. During that limited time, students must assemble their experimental apparatus, perform the experiment, then dismantle and clean their apparatus. In some cases, they have as little as ninety minutes for their laboratory period. Second, to be meaningful, such an experience needs to be reflective of the courses they are taking and have taken.

At the University of Houston, we have refashioned the Junior laboratory to meet these parameters. In this presentation, we outline how we devised, designed, tested, and implemented a set of experiments that Junior chemical engineering students can do quickly and efficiently. We also discuss how we integrated the laboratory and course work so that the student could directly apply classroom learning to evaluating experimental data.

(150g) Using Molecular Simulation to Improve the Efficiency of Biomass Processing

Jarin, Z.,

Originally presented on: 11/4/2013 14:30:00 - 14:50:00

Interest in biomass as a feedstock for chemical processes has been increasing over the last several years. In order to compete economically, conversion of biomass to small molecules for production of fuels and value-added chemicals must become more efficient. However, a major hurdle to chemical processing is the robust structure of biomass and subsequent resistance to breakdown. In the past ten years, experiments have shown ionic liquids, a special group of solvents, have the rare ability to dissolve biomass. While there are countless ionic liquids known to the research community, these experiments are not guided by molecular scale design principles, hindering more promising research. This project focuses on providing fundamental knowledge to explain the ionic liquid-mediated biomass dissolution by using molecular dynamics (MD) coupled with enhanced sampling techniques. Enhanced sampling allows the efficient exploration of the free energy landscape of a system by modifying MD algorithms. Specifically, an enhanced sampling technique called metadynamics will be used to explore glucose's conformational free energy surface, which can be used to better understand the mechanism of dissolution in ionic liquids. Insight from these simulations can then be applied to larger sugar oligomers that constitute cellulosic biomass.

(210a) Rapid Skin Permeabilization By The Simultaneous Application Of Dual-Frequency Ultrasound

Maa, R.,

Originally presented on: 11/4/2013 15:15:00 - 15:35:00

Transdermal drug delivery is an attractive delivery route because of the ease of access of the skin. However, the skin acts as an effective barrier. Therefore, large molecules must be delivered with needles, which are painful and can increase the transmission of infectious diseases. Low frequency ultrasound has previously been studied for its ability to painlessly increase the permeability of the skin. Despite this work, ultrasound has not yet attained wide clinical use due to limitations on the resulting permeability. We hypothesized that skin permeability can be enhanced by coupling a high-frequency ultrasound signal to the existing low-frequency treatment. Aluminum foil pitting experiments, along with in vitro tests performed with porcine skin, tested the simultaneous application of two distinct frequencies ranging from 20 kHz to 3 MHz to examine their effect on the formation of localized transport regions (LTRs) within the skin. These experiments demonstrated that skin permeability and resulting LTR size both increased as a result of dual-frequency treatment compared to low-frequency treatment alone. Glucose (180 Da) and inulin (5000 Da) transdermal flux experiments further confirmed these results by demonstrating enhanced delivery through and into the dual-frequency-treated skin. This methodology can be a viable option for transdermal drug delivery.

(210b) N/a

Zeidman, J., N/a

Originally presented on: 11/4/2013 15:35:00 - 15:55:00

The controlled production of artificial antibodies for therapeutic treatments is a rapidly growing and profitable field. One primary method of antibody production is transient transfection in mammalian cells most often using polyethyleneimine (PEI) as the transfection agent and Human Embryonic Kidney (HEK) or Chinese Hamster Ovary (CHO) stable cell lines. Despite its importance, transfection is not well understood and leads to variable results based on many factors including medium used, transfection and incubation times, viable cell density (VCD) at inoculation, PEI concentration, and DNA concentration. The quantity and quality of antibody produced after transfection are of the upmost importance. In this experiment, a specific transfection protocol was optimized for antibody titer in HEK cells. Design of Experiments software was utilized to plan experiments to maximize antibody titer by varying the parameters of inoculation VCD, PEI concentration, and transfection time. Experiments were carried out in a six-chamber HexaScreen® mini-bioreactor with a control and scale-up performed with a 2 liter roller bottle system. Using the HexaScreen, 15 different trials were performed over the given parameters to conclude that the optimal combination was an inoculation VCD of 3 million cells/mL, a PEI concentration of 35 mg/L, and no transfection time, which would yield an antibody concentration over 40 mg/L. The roller bottle scale-up demonstrated inconsistency and scale-up problems that would need to be addressed with future experiments. While further experiments are necessary, this project demonstrated the feasibility of this system for better understanding and optimizing transient transfection and some of the relations between antibody titer and transfection parameters.

(210c) A Novel Formulation For Non-Swelling Peg Hydrogels For Application In Spinal Cord Injury (SCI)

Fu, M.,

Originally presented on: 11/4/2013 15:55:00 - 16:15:00

The notion of using hydrogels, which are 3-dimensional cross-linked polymer networks, for the delivery of biologically relevant molecules has been around for decades. However, with regard to administering hydrogels to anatomically sensitive areas like the spinal cord, there still exist 3 main challenges: the mechanical stress conferred by the swelling of the hydrogel, the rate of degradation of the gel network to control drug release, and the capacity of the hydrogel to maintain the efficacy of the drug over time. In this project, all three areas are addressed. Testing of various novel formulations for PEG hydrogels has yielded a network that exhibits temperature-sensitive syneresis, which allows for complete control of swelling. In addition, specific additional modifications of the chemistry to alter the gel matrix's susceptibility to hydrolysis has allowed for tailoring of the degradation of the hydrogel. As a result, precise tailoring of therapeutic drug release rate from the gel is attainable. Finally, the efficacy of incorporating stabilizing excipients into the gel network as a method of preserving the enzymatic activity of therapeutic proteins is being investigated and has shown tremendous preliminary potential. In particular, the incorporation of trehalose has been shown to confer stabilizing properties to the hydrogel as a whole with respect to the enzyme Chondroitinase ABC, which has therapeutic applications in spinal cord injuries.

(210d) The Quantification Of Mitochondrial Location and Morphology Using Matlab

Bussie, B., AUBURN UNIVERSITY
Kerscher, P.,
Lipke, E. A., Auburn University
Originally presented on: 11/4/2013 16:15:00 - 16:35:00

The analysis and quantification of biological images is essential for researchers in many disciplines. For proper statistical analysis and results justification, hundreds of cells frequently have to be analyzed by manual counting due to the limitations of image processing tools. In order to better understand the function of human cells and tissues and to be able to use that knowledge to both prevent and treat disease, we need to be able to quantify visual information contained in biological images accurately in a time-efficient, reproducible, and unbiased manner.

A MATLAB program was created to analyze fluorescently labeled organelles within cells. Specifically, the program was used to locate and quantify mitochondria and nuclei of different cell types, including human induced pluripotent stem cells (hiPSC), human embryonic kidney cells (HEK-293), and a human breast cancer cell line (MCF-7). The program extracts information about the number of mitochondria within the cell, as well as location of the mitochondria relative to the nucleus. Information about the morphology of the mitochondria, such as shape and connectivity was also determined. Compared with manual interpretation, it was found that the MATLAB program was quicker, more consistent, and bias-free. When applied to differentiating stem cells, this program can reveal how the organization and morphology of mitochondria changes with cell maturation and differentiation.

(210e) Analysis Of CELL Proliferation Within A 3D Tissue-Engineered Cancer MODEL

Clary, J.,
Pradhan, S., Auburn University
Lipke, E. A., Auburn University
Originally presented on: 11/4/2013 16:35:00 - 16:55:00

Breast cancer is one of the most common forms of cancer throughout the world. Researchers are attempting to develop an artificial platform for growing cancer cells outside the human body for the study of the disease. The current study aims to create an *in vitro* environment that accurately mimics tumorigenic behavior. To accomplish this, human breast cancer MCF7 cells were encapsulated in polymer hydrogel microspheres composed of poly(ethylene glycol) and fibrinogen (PEG-Fb) using an oil-in-water emulsion technique and maintained in culture long-term. The microspheres were analyzed for size changes using ImageJ. SEM imaging was conducted on the microspheres to visualize the distribution of cells within the polymer matrix. Additionally, the MCF7 cells were stained with Hoechst 33342 dye after enzymatic-dissociation of the hydrogel microspheres for measurement of cell proliferation using the total DNA assay. The results obtained from the analysis indicated that total DNA content increased and the cells proliferated within the polymer hydrogel system over time. Thus, an artificial 3D platform for long-term culturing of cancer cells has been established. Further biochemical assays on the developed system are necessary to make it more closely mimic native breast cancer conditions so that cancer therapies may be reliably studied *in vitro*.

(210f) Membrane Compatibility With Switchable Polarity Draw Solutions For Use In Forward Osmosis Applications

Coscia, B. J., University of Connecticut
Reimund, K. K., University of Connecticut
Wilson, A. D., Idaho National Laboratory
McCutcheon, J. R., University of Connecticut
Originally presented on: 11/4/2013 16:55:00 - 17:15:00

Water scarcity is problem being faced worldwide and present in every continent. Close to one-fifth of the world's population has difficulty acquiring safe water, and the problem is worsening as populations continue to grow in poorer countries. As the availability of unimpaired freshwater sources dwindles, cew water sources, such as the oceans and saline groundwaters, must be tapped. However, desalination technologies are very expensive due to a high energy requirement. Forward osmosis (FO) is a process which may be able to replace or become integrated with existing desalination technologies like reverse osmosis. FO relies on an osmotic agent, or draw solution, to drive water from a saline water source by osmosis. The water dilutes the draw solution and a secondary separation process where the solute is removed and recycled. This secondary separation step is key to FO process economics. Recently, a classification of draw solutes, known as switchable polarity solvents (SPSs) have emerged. These solvents become highly miscible with water upon sparging with CO_2 and phase separate when stripped with an inert gas. One such SPS is N,N-dimethylcyclohexylamine. This SPS has shown promise in early testing at Idaho National Laboratories. However, because it is a solvent, membranes, which are polymeric, may not be compatible with the solution. . This study examines the longevity of conventional osmotic membranes after exposure to relevant SPS solutions. Membrane performance metrics, such as permeance and salt rejection as well as osmotic flux, are presented.

Force to Death and New Life, Some with Programming Vitality

Papoutsakis, E. T., University of Delaware

Originally presented on: 11/4/2013 18:00:00 - 18:25:00

Regenerative medicine is the quest to provide biological or hybrid synthetic-biological materials that can enable or facilitate therapeutic interventions, from transfusions and gene therapy to organ replacement. A long-standing goal is to ability to produce human blood cells for transfusion medicine. Among blood cells, platelets, needed for blood coagulation and vascular repair, are an expensive "product" in limited supply. This is due to the collection and processing steps from donated blood, and the fact that platelets cannot be frozen, but also due to the possibility of bacterial or blood-borne pathogen contamination. Can we produce platelets in a "blood factory"? This is recognized as a grand challenge that remains largely elusive. Platelets are produced from the large, polyploidy megakaryocytes (Mks) in the bone marrow and the lung vasculature, under a spectrum of biomechanical forces. I will discuss these forces and show how important they are for producing functional proplatelets and other, small, anuclear particles, which we classify as Mk microparticles (MkMPs). I will show the extraordinary ability of these MkMPs in programming stem and progenitor cells, and discuss their potential as a means for targeted delivery of nucleic acids and proteins. I will also discuss our explorations aiming to understand the mechanisms by which MkMPs act, and will argue for producing and using them for regenerative-medicine applications.

(248a) Synthesis of Fine Chemicals by Heterogeneous Catalysis: A Critical Approach of the Role of Solvent for Hydrogenation Reactions

Apesteguia, C.,

Originally presented on: 11/5/2013 8:30:00 - 8:55:00

The synthesis of fine chemicals over solid catalysts often involves the use of solvents that may strongly influence the catalyst performance. Thus, the choice of suitable solvents is frequently critical to obtain high catalytic activity and selectivity. However, the optimal solvent selection requires a detailed knowledge on the relationship between the chemical nature of the solvents and the interactions taking place in the gas-liquid-solid catalytic systems. One of the most common types of catalytic reactions carried out in the presence of solvents is the hydrogenation of organic compounds. In particular, the selective hydrogenation of aromatic ketones into the corresponding alcohols on metal-based catalysts in the presence of different solvents has been widely studied. Here, the solvent effect on catalyst activity and selectivity for the liquid-phase hydrogenation of acetophenone to 1-phenylethanol was thoroughly investigated over Ni/SiO_2. Solvents of different properties and polarities were used: i) protic solvents: methanol, ethanol, 1-propanol and 2-propanol; ii) aprotic polar solvents: acetonitrile, g-butyrolactone and tetrahydrofuran; iii) aprotic apolar solvents: cyclohexane, toluene and benzene. The relative interactions solvent–catalyst, solvent–reactant and reactant–catalyst and their influence on the activity pattern were considered in the analysis. Concerning the influence of the solvent-reactant interaction, classical polarity parameters (e.g. dipole moment m and dielectric constant ε) and other solvatochromic scales (e.g. hydrogen-bond donor (a) and hydrogen-bond acceptor (b) parameters, p* polarity/polarizability index, Kosower's Z and ET(30) scales) were taken into account. The influence of H_2 solubility in the solvents was also considered. For the solvent-catalyst and reactant-catalyst interactions, the corresponding molar adsorption enthalpies were measured calorimetrically and compared.

(267b) Career Development: It Ain't A Random Walk

Worstell, J. H., Worstell and Worstell, Consultants

Most chemical engineers do not realize the difference between a job and a career. A job is a forty year random walk through the chemical processing industry. A career is a directed path to an ultimate goal. There is a second feature about careers that chemical engineers do not recognize: careers begin with a written plan. But, to write a career development plan, the chemical engineer must understand corporate structure. This presentation outlines corporate structure in terms of leadership; i.e., direct leadership, organization leadership, and strategic leadership. It then shows how to use this structure to formulate, develop, and manage a career in the chemical processing industry. This presentation also discusses the content of such a career plan and the need for it to contain three levels based on a chemical engineer's twenty year goals, which dictates the ten year goals that in turn dictate the five year goals. This presentation also discusses the need for such a plan to be evergreen.

(267a) An Industry Career in Process Development: Leveraging Chemical Engineering Training

Powell, J. B., Shell Projects & Technology

Process development is a key skill for industry, bridging from innovation to commercialization. Skills required span the universe of chemical engineering. The broad exposure provides a path to lifelong learning, and upward mobility in the organization. This poster will present an overview of process development as a career path, and discuss key success factors for managing a process development career.

(267d) Chance Favors The Prepared Mind

Hill, M., Columbia University

The author's chemical engineering career has followed a tortuous path, including episodes as a college dropout, a research engineer for a soap company, a strategic leader in the consumer products industry, and a lecturer at an Ivy League university. He will share some anecdotes from his career that illustrate important lessons for today's young professionals.

(281a) At Least 1,000X Thinner than a Human Hair

Korgel, B. A., The University of Texas at Austin

Originally presented on: 11/5/2013 11:15:00 - 11:40:00

The last century witnessed the development of two classes of materials—polymers and semiconductors—that dramatically changed people's lives. A new class of materials—nanomaterials—has emerged with the potential to create a wide range of new technologies because of their unique combination of properties. For example, semiconductor nanowires have attributes of polymers like mechanical flexibility, light-weight and low cost, combined with the useful electronic and optical properties of semiconductors into one material. Nanowires can be embedded in polymer hosts, spun into fibers, or dispersed in solvents and printed on substrates like organic materials. They might be utilized to create functional textiles with unprecedented capability, such as fabrics with impressive structural integrity capable of generating power from the sun as in a photovoltaic device or store energy in the form of a battery. Nanocrystals of various semiconductors can also be made and formulated into paint that can be used to deposit semiconductor films at low cost onto large-area, mechanically flexible and light-weight substrates for a next generation of low-cost solar cells. This presentation will highlight nanomaterials we have been studying for lithium ion battery and photovoltaics applications in particular and some of the challenges facing commercial use.

(335a) Pot Growth Of Sweet Corn On Acidic Soil Ameliorated By a Modified Waste Mud In Guangdong Province, P R China

Shi, L., Iowa State University
Originally presented on: 11/5/2013 12:30:00 - 12:48:00

Pot growth of sweet corn on acidic soil ameliorated by a modified waste mud in Guangdong Province, P R China

Shi Lin Celshi@scut.edu.cn

South China University of Technology, Guangzhou 510640, P. R. China

Abstract: A modified waste mud was prepared by calcing a mixture of soda waste mud and potassium feldspar. The nutrient contents in the modified waste mud dissolved in 0.1 mol·L^{-1} citric acid solution were 23.43% CaO, 22.92% SiO_2, 4.27% K_2O, 5.04% MgO. Its pH value held 9.75. The contents of heavy metals in the modified waste mud were much lower than the ecological index of lead, cadmium, chromium and mercury for fertilizers (GB/T 23349-2009) and background value in soil except for arsenic. In this study, nine different treatments with addition of the raw waste mud 1,2,3,4 g per kilogram soil (LW_{1-4}) and the modified waste mud 1,2,3,4 g per kilogram soil (LM_{1-4}) respectively and a control (CK) were planted with sweet corn for 30 days. All the planting tests were repeated for 4 pots. The results indicated that applying the modified waste mud had better effects than adding the raw waste mud only. Compared with CK, the seedling emergence of sweet corn, the plant height and stem thickness, the dry and wet biomass achieved 33.33%~8.34%, 28.38%, 32.20%, 70.51% and 80.10% respectively. With the addition of the modified waste mud, the content of available potassium, exchangeable calcium and exchangeable magnesium in the planted acidic soil was higher than that of CK, which was beneficial to plant growth. Taking the rate of sweet corn seedling emergence, the plant height and stem thickness, the dry and wet biomass into account, the optimal application rate of the modified waste mud was 2g per kilogram soil, which could bring forward the growth of sweet corn evidently and ameliorate acidic soil, distributed widely Guangdong, P R China.

(335b) The Photocatalytic Activity Of ZnO/ZnS Core/Shell Particles For Methyl Orange
Li, W.,
Originally presented on: 11/5/2013 12:48:00 - 13:06:00

The photocatalytic activity of ZnO/ZnS core/shell particles for methyl orange

Wenjiang Li[*a], Ge Song[a], Minfang Chen[a]

Abstract: Recently, many novel properties of ZnO have been developed not only by the impurity doping but also through the surface modification [1]. A heterostructure formed by ZnO and another semiconductor can present a highly efficient photoelectric effect in the solar cells [2,3]. As is well known, ZnS with wide band gap also attracts much interest in photocatalysts, photoconductors, optical sensors, electroluminescent materials [12]. The band gap of ZnS (3.68eV) is larger than that of ZnO, and theoretical calculations and experimental results have demonstrated that the combination of these two semiconductors could yield a novel composite with the photo excitation threshold energy lower than those of the individual components [4]. In this paper, the novel spherical ZnO/ZnS core/shell particles were obtained after sulfurizing ZnO spheres by Na_2S. The pure spherical ZnO particles were firstly synthesized via a simple mild solvothermal approach by the assistance of diethylene glycol. Then, the partial ZnO molecules on the surface of the ZnO spheres were converted to ZnS molecules in the presence of S^{2-} ions under hydrothermal condition, and finally forming ZnS shell structure over the surface of ZnO spheres. The thickness of the ZnS shell is adjustable by controlling the sulfuration process of as-synthesized ZnO template. The photocatalytic activity of the ZnO/ZnS core/shell composite was evaluated by monitoring methyl orange degradation as the probe reaction.

(335c) Biosynthesis of Polyhydroxybutyrate By Thermotolerant Chelatococcus Daeguensis Strain TAD1 At Elevated Temperatures: A Potential Low-Cost Way for Large-Scale Production

Huang, S.,

Originally presented on: 11/5/2013 13:06:00 - 13:24:00

Biosynthesis of polyhydroxybutyrate by thermotolerant *Chelatococcus daeguensis* strain TAD1 at elevated temperatures: A potential low-cost way for large-scale production

Shaobin Huang[a,b,c,*], Fuqian Xu[a,c], Bichun Huang[a], Ruimin Huang[a], Yongwen Ma[c,d]

[a]College of Environment and Energy, South China University of Technology, Higher Education Mega Center, Guangzhou 510006, P.R. China

[b]State Key Laboratory of Pulp and Paper Engineering, South China University of Technology, Guangzhou 510640, P.R. China

* Corresponding author. Tel/Fax: 86-020-39380587 Email: chshuang@scut.edu.cn

Abstract: This study set out to investigate the polyhydroxybutyrate (PHB) accumulation properties of a newly isolated thermophilic denitrifier-*Chelatococcus daeguensis* TAD1. TAD1 could accumulate PHB effectively at a broad range of temperatures (from 35 °C to 55 °C), and the highest cell density and PHB-accumulation were 3.6 g cell dry weight (CDW) per liter and 83.6% of CDW (w/w) at 45 °C, respectively. Although lower nitrogen concentrations showed higher PHB accumulation, the optimal PHB production of TAD1 was acquired at a medium concentration of 311.13±3.7 mg/L NH_4^+-N (C/N=30), where the cell density and PHB content were 3.86 g CDW per liter and 84.1%, respectively. Other inexpensive carbon sources were evaluated from the perspective of product yield ($Y_{P/S}$, gram PHB per gram substrate used), and the glycerol obtained the best results. Moreover, TAD1 could also utilize starch to produce PHB at elevated temperatures, which has rarely been reported by other microorganisms.

(335d) Effects Of Phenol On The Degradation Of Reactive Brilliant Blue Kn-R In SR-AOP

Ma, Y.,

Originally presented on: 11/5/2013 13:24:00 - 13:42:00

Effects of phenol on the degradation of Reactive brilliant Blue KN-R in SR-AOP

Yongwen Ma [a, b, c], Jinquan Wan[a, b, c], Yan Wang [a], Mingzhi Huang [a], XinGuo[b]

[a] *College of Environmental Science and Engineering, South China University of Technology, Guangzhou 510006, PR China*

[b] *State Key Laboratory of Pulp and Paper Engineering, South China University of Technology, Guangzhou, 510640, China*

[c] *The Key Lab of Pollution Control and Ecosystem Restoration in Industry Clusters, Ministry of Education, South China University of Technology, Guangzhou, 510006, China*

Abstract: Simulation wastewater mixed with Phenol and Reactive Brilliant Blue KN-R was treated by Sulfate radical-based advanced oxidation process. The competitioneffects of two kinds of substrate were studied through a series of batch experiments. The experiment results showed thatthe degradation rate of Reactive Brilliant Blue KN-R increased with the increase of the ferrous ion and persulfate concentration in the oxidation system only Reactive Brilliant Blue KN-R existed. The process of sulfate radical oxidation could be divided into two regions according to the secondary kinetic characteristics. When the phenol was added to solution, the degradation rate of Reactive Brilliant Blue KN-R was decrease with increase of phenol concentration; the phenol consumed sulfate radical competed with Reactive Brilliant Blue KN-R. The process of oxidation also divided into two regions, and the two parts according to the secondary kinetic characteristics respectively, with reaction rate constant was decrease as increase of phenol concentration.

(335e) Effects Of Oxygen Plasma Pretreatment Of Mwcnts In MnOx/Mwcnts Catalyst For Selective Catalytic Reduction Of NO

Huang, B.,

Originally presented on: 11/5/2013 13:42:00 - 14:00:00

Effects of oxygen plasma pretreatment of MWCNTs in MnOx/MWCNTs catalyst for selective catalytic reduction of NO

Bichun Huang[a,b] *, Wenhui Wang[a], Lishan Wang[a]

[a] *College of Environment and Energy, South China University of Technology, Guangzhou, 510006, China*

[b] *Guangdong Provincial Key Laboratory of Atmospheric Environment and Pollution Control, Guangzhou, 510006, China*

* Corresponding author. Tel: +86 20 39380519; E-mail: cebhuang@scut.edu.cn (B. Huang)

Abstract

Multi-walled carbon nanotubes (MWCNTs) have become promising support materials for catalysis due to their high surface areas, structural integrity, and controllable dimensional characteristics[1-2]. However, the application remains limited for the inert nature of the primitive MWCNTs surface. Pretreatment with oxygen plasma will produce oxygen-containing groups on the surface of MWCNTs, which can act as effective sites in selective catalytic reduction (SCR)[3].

Tab. 1 Pore structure parameters of the catalyst

Sample	S_{BET} (m^2/g)	V_{total} (cm^3/g)	Pore size (nm)
MnOx/MWCNTs	79	0.213	10.7
MnOx/MWCNTs-20	89	0.236	10.6
MnOx/MWCNTs-40	95	0.257	10.8

Fig. 1 The effect of modification time on NO conversion

In this research, MWCNTs were pretreated in oxygen dielectric barrier discharge (DBD) plasma with power of 15W and a fixed frequency 50Hz for 0, 20, 40 min. Then, a series of MnOx/MWCNTs catalysts were prepared by conventional wet impregnation method. The samples obtained were denoted as follows: MnOx/MWCNTs (without oxygen DBD plasma pretreatment), MnOx/MWCNTs-20 (pretreated in plasma for 20 min) and MnOx/MWCNTs-40

(pretreated in plasma for 40 min).The catalytic properties in NH_3-SCR for these catalysts were compared. The catalytic activity was carried out in a quartz fixed-bed continuous flow reactor. The MnOx/MWCNTs catalysts were characterized by Brunauer-Emmett-Teller (BET) surface area, X-ray powder diffraction (XRD) and NO temperature programmed desorption (NO-TPD). Contributed to the more oxygen-containing groups, the catalysts supported on the plasma modified MWCNTs had better catalytic activity and wider temperature windows (Fig. 1). These catalysts had higher surface area (Tab. 1), better Mn dispersion (Fig. 2) and larger NO absorption amount (Fig. 3) compared with the catalyst supported on unmodified MWCNTs. Besides, the catalytic properties strengthened as a function of discharge time.

Fig. 2 XRD spectrum of different MnOx/WMCNTs catalysts

Fig. 3 NO-TPD profile of different MnOx/WMCNTs catalyst

Furthermore, the MnOx/MWCNTs-40 sample was used as the representation of catalyst to study the mechanism of SCR. DRIFTS results revealed that the E-R and L-H mechanism both occurred on the surface of the catalyst. However, the former was the dominant pathway.

Keywords: MnOx/MWCNTs; oxygen DBD plasma; selective catalytic reduction (SCR)

*** Formatting Issue - Best Available Paper/Graphic ***

(335f) Adsorption and Photodegradation Of Dimethyl Phthalate In An Aqueous Solution By TiO2/ Hydrotalcite Composites
Wu, P.,
Originally presented on: 11/5/2013 14:00:00 - 14:18:00

Adsorption and Photodegradation of Dimethyl Phthalate in an Aqueous Solution by TiO_2/ Hydrotalcite Composites

Pingxiao Wu*[1], Zhujian Huang[1]

[1]College of Environment and Energy, South China University of Technology;

*pppxwu@scut.edu.cn

Environmental endocrine phthalates (PAEs) is a hydrophobic trace organic pollutant commonly existed in aquatic environment. Dimethyl phthalate (DMP) is one of the relatively stable compounds in natural environment, and its hydrolysis half-life is estimated to be about 20 years. Photocatalytic oxidation is a commonly used, advanced oxidation technology, which can effectively degrade organic pollutants. The commercial TiO_2 nanoparticles have many disadvantages for photocatalytic degrading hydrophobic trace organic pollutants. Hydrotalcite have excellent properties and are an appropriate disperser and carrier for TiO_2 nanoparticles. This paper presents a new organic hydrotalcite/TiO_2 composite in an improved preparation method to efficiently remove DMP in an aqueous solution. The hydrotalcite:TiO_2 weight ratio = 4:1, 4:2, 4:3, 4:4, 3:4, 2:4 was labeled as HT41, HT42, HT43, HT44, HT34, and HT24, respectively. The optimal hydrophobic hydrotalcite/nano-TiO_2 ratio was screened by the photocatalytic experiments of DMP, and the photocatalytic mechanism was discussed. This study aims at providing technology and theory for purifying drinking water contaminated by organic pollutant with low concentration.

The result of XRD indicated that sodium dodecyl sulfate (SDS) anions are successfully intercalated into the interlayer of hydrotalcite. The arrangement of SDS anions in the gallery of hydrotalcite is confirmed by 1-D electron density projection. The orientation angle between the SDS anions and c-axis is calculated to be 30°. As the ratio of TiO_2 to organic hydrotalcite increases in the composites, randomly immobilized TiO_2 onto the surface of high-ordered sheets forms a mixed and disordered composite. High resolution Ti 2p and O 1s XPS showed that TiO_2 dispersed on the SDS-hydrotalcite still has octahedrally coordinated anatase

*** Formatting Issue - Best Available Paper/Graphic ***

structure. Also, TiO_2 and SDS-hydrotalcite are not simply a physical mix, but combined together by chemical bonds. SEM image (Fig. 1) provides direct evidence that the reconstructed SDS-hydrotalcite are consisted of interconnecting nano-flakes with an average thickness of approximately 20 nm, which would provide abundant adsorption sites for adsorbate. The results (Fig. 2) of adsorption and photodegradation experiment showed that SDS-hydrotalcite/TiO_2 composites can effectively adsorb DMP in aqueous solution, which will promote the catalysts photocatalytic degrade the adsorbates in situ. Both adsorption and photocatalytic degradation contribute to removal of DMP with SDS-hydrotalcite/TiO_2 composites under UV light. The photocatalytic ability of SDS-hydrotalcite/TiO_2 composites is much higher than that of bare TiO_2. The optimum mass ratio of hydrotalcite: TiO_2 is 1:1. HR-XPS patterns of fresh and used SDS-hydrotalcite/TiO_2 composites indicated that the photogenerated electrons incline to transfer onto the surface of Mg in the hydrotalcite. Such an electron transfer leads to the depression of electron-hole recombination and the improvement of the photocatalytic activity of TiO_2.

Fig. 1 SEM image of TiO_2/hydrotalcite composites

Fig. 2 Removal of DMP by TiO_2/hydrotalcite composites.

In conclusion, the photocatalytic properties of commercial TiO_2 nanoparticles for degrading hydrophobic pollutant DMP could be effectively enhanced by a simple method through hybridizing them with the interconnecting SDS-hydrotalcite nano-flakes, and SDS-hydrotalcite/TiO_2 composites are confirmed to combine together by chemical bonds. The optimum amount of the hydrotalcite powders as a carrier for the enhancement of the photocatalytic reactivity of the SDS-hydrotalcite/TiO_2 system was estimated to be the mass ratio of 1:1. The SDS-hydrotalcite/TiO_2 composites showed higher photocatalytic reactivity than that of bare TiO_2 nanoparticles mainly because SDS-hydrotalcite with flaky structure worked as an effective adsorbent to condense DMP in the aqueous solution.

(335g) Comprehensive Utilization of Selenium Resources in Black Shales of Enshi
Shuai, Q.,
Originally presented on: 11/5/2013 14:18:00 - 14:36:00

Comprehensive utilization of selenium resources in black shales of Enshi

Qin Shuai[1*], Shengrui Xu[1], Yu Tang[1], Yunjie Huang[1], Sen Yan[2], Zhengyu BaO[1,2]

[1] *Faculty of Material Science and Chemistry, China University of Geosciences, Wuhan, 430074, China*

[2] *Faculty of Earth Sciences, China University of Geosciences, Wuhan, 430074, China*

Selenium (Se), as an important industrial material, has been extensively applied in many fields such as electronic metallurgical and petroleum industry.[1] However, selenium is a typical dispersed element with a low crustal concentration (0.05-0.09 μg g^{-1}). Interestingly, the huge independent deposit of selenium with 5 billion tons of selenium resources has been found in Enshi region (Hubei Province, China) and thought as the only site in the world till now. [2] The selenium with high concentration of more than 80000 μg g^{-1} locates in the black shale which is called stone coal. [3] The stone coal is popularly utilized through combustion process. By this way, a significant portion of selenium in the vapors will escape into atmosphere and then lead to air pollution and waste of the scarce resources. Therefore, it is significant to increase the comprehensive utilization efficiency of selenium in the Se-rich stone coal during the combustion process.

Our work was focused on the adsorption and recovery of Se in the vapors from Se-rich stone coal combustion processes. Various calcium-based sorbents were firstly developed and evaluated for the Se adsorption at 700°C-1000°C. It was found that adsorption efficiency of selenium could reach 90.6% at 800°C with CaO/nano-ZnO as sorbent. This adsorption effect was from the formation of $CaSeO_4$ species. Then, the recovery of the Se from $CaSeO_4$ species were also studied with the reducing agent of Na_2SO_3. High percents of 98.17% and 69.78% were obtained for the reduction of Se from simulation and real experiments, respectively. The developed adsorption and recovery methods had potentials to be used for the utilization of Se resources from stone coal.

(335h) Imbibition Oil Recovery Theory and Influencing Factors
Zhang, Y.,
Originally presented on: 11/5/2013 14:36:00 - 14:54:00

Imbibition Oil Recovery Theory and Influencing Factors

Zhang Yi [1], Han Da-kuang [1], Wu Kang-Yun[1] ,Rao Huan[2]

1. State Key Laboratory of Enhanced Oil Recovery (Research Institute of Petroleum Exploration and Development), Beijing 100083, China

2. Institute of chemistry Chinese Academy of Scineces, Beijing 100190, China

AbstractÂ □ F

Imbibition oil recovery theory main includes capillary forces, wettability alteration, ion desorption and flow theory. There are many effects on imbibition, including the reservoir rock wettability, reservoir properties, fluid properties and reservoir environmental. However, the scholars' views are not consistent. The effect factors and the relationship for imbibition oil recovery are comparative analysis, and the latest experimental research methods are introduced. The future direction of development of the imbibition oil recovery theory and the problem need to solve were predicted.

(335i) Developing the Green Chemistry Principles

Cai, W., Wuhan University of Technology
Originally presented on: 11/5/2013 14:54:00 - 15:12:00

Developing the Green Chemistry Principles

Weiquan Cai

School of Chemical Engineering, Wuhan University of Technology, Luoshi Road 205#, Wuhan 430070, P. R. China
E-mail address: caiwq@whut.edu.cn, wqcai@home.ipe.ac.cn

To reduce or eliminate chemical pollution in origin is the ideal of green chemistry, while its principles are the best elucidation for its essence. Based on the 12 wellknown principles proposed by Anastas et al in 1998, the 12 more principles were complemented by Anastas himself and Winterton, respectively from technical, economic and commercial considerations. Unfortunately, these principles don't explicitly include concept which is highly relevant to environmental impact. As a result, the concept of green engineering and its principles were proposed. Tang et al subsequently simplified the original 12 green chemistry and 12 green engineering principles proposed by Anastas et al as the mnemonic, communicative and popularized "IMPROVEMENTS PRODUCTIVELY". However, the principles are still intuitive and commonsensible and thus can't explicitly reflect the inherent relationships between the goal and related research fields of green chemistry. It is estimated that the green chemistry principles which are centered around designation of green, coordination of Atom Economy and Effective Mass Yield etc., data collection of life-cycle assessment, integration of properties of catalyst with reaction, separation and recycling use, and the optimization and balance of evaluation metrics and methods to greenness etc. still need to be developed and upgraded over a long period of time.

(335j) Synthesis Technics of Aliphatic Dicarboxylic Acid Alkoxy Ethyl Ester
Chen, M.,
Originally presented on: 11/5/2013 15:12:00 - 15:30:00

Synthesis Technics of Aliphatic dicarboxylic acid alkoxy ethyl ester

CHEN Muhua[1,2], *WANG Junxiang*[1],*ZHANG Xiaoxiang*[1], *ZHU Xinbao*[1,2]

(1. College of Chemical Engineering,Nanjing Forestry University,Nanjing 210037,ChinaÂ
□ G2.Engineering Research Center for Alcohol Ether of Jiangsu Province,Wuxi 214441,China)

CHEN Muhua :simonzrh@163.com

AbstractÂ □ F Aliphatic dicarboxylic acid alkoxy ethyl ester is a kind of cold resistant plasticizer with good performances. This kind of dibasic ester has high plasticizing effectiveness with PVC and PVA, and provides excellent low temperature flexibility, a certain light stability and water resistance. So it has been widely used in nitrile rubber,polychloroprene, acrylic rubber, polyvinyl chloride,cellulose acetate resin,and other fields. In the present research, Aliphatic dicarboxylic acid ester was synthesized from aliphatic dicarboxylic acid and ethyle glycol monoalkyl ether,by using hetergeneous acid as the catalyst, butyl ether as the dehydrant. Followed by filtration and distillation, the high purity product was obtained easily. Some commercial (ZSM-5 and HZSM-5) and homemade catalysts (SO_4^{2-}/AC and SO_4^{2-}/TiO_2-ZrO_2) were investigated by esterification and characterized using gas adsorption analysis (BET), X-ray diffraction (XRD), X-ray fluorescence (XRF), temperature-programmed decomposition (TPD) and electron spin resonance (ESR) techniques. The weight-based activity of the heterogeneous catalysts decreases in the following order: SO_4^{2-}/TiO_2-ZrO_2>SO_4^{2-}/AC>HZSM-5≈ZSM-5. It was unexpected that the esterification rate of ZSM-5 was up to 92% and the good catalytic activity was maintained during recycling tests. According to the characterization results, it was revealed that the exchange between hydrogen ion (from aliphatic dicarboxylic acid) and sodium ion (on the catalyst surface) plays an important role during the reaction. Upon optimizing the catalyst preparation conditions and reaction parameters, both SO_4^{2-}/TiO_2-ZrO_2 and SO_4^{2-}/AC exhibited 99% initial esterification rate. For SO_4^{2-}/TiO_2-ZrO_2, it was possible to recycle the catalyst without visible loss of activity because of the big amount of the acid sites and its strange acidity on the catalyst surface. However, the results also indicated that loss of sulfonic acid groups and pore blockage in activated carbon channels was two main reasons for catalyst deactivation. In addition, apparent rate constants for those acid-catalyzed homogeneous esterification reactions were obtained,and apparent kinetics models were established. The results showed that the synthesis of dibasic esters (between 453.15-473.15K) was the third order irreversible reaction. Activation energy values (*E*a) of those solid catalysts were all in the range from 110-140 kJ/mol, which was obviously lower than comparative data in homogeneous reactions (220 to 250 kJ/mol).

(335k) Research On the Performance of SCR of NOx of Loaded Catalysts With Activated Carbon As Carrier

Gui, K., Xiamen University
Originally presented on: 11/5/2013 15:30:00 - 15:48:00

Research on the Performance of SCR of NOx of Loaded Catalysts with Activated Carbon as Carrier

Gui Keting Wu Haimiao Wang Xiaobo

Key Laboatoy of Enegy Themal Convesion and Contol of Ministy of Education, School of Enegy and Envionment, Southeast Univesity, Nanjing, 210096

Gui Keting:101000477@seu.edu.cn

Abstract: There are many ways to control NO_x emissions, and the selective catalytic reduction (SCR) is the most widely used method. It has relatively high removal efficiency, and consumes less reducing agent. But in the SCR of NO_x process, catalyst is expensive, the high temperature of catalyst active window causes the flue gas reheating, and wasted catalyst will lead to secondary pollution. Therefore, it is very significant to develop new efficient and low-temperature catalysts. Hence, the new loaded catalyst was prepared through choosing activated carbon as carrier, and the transition metal (FeÂ □ AMnÂ □ ACuÂ □ ACr) as loaded materials in the paper. Some parameters, such as reaction components, loaded quantity, reaction temperature, oxygen concentration and superficial velocity on the efficiency of SCR of NO_x were experimental studied. The results show that catalyst 8Fe-8Mn/AC exhibites high catalytic activity and is a good low-temperature catalyst.

(335l) Determining Stoichiometry in p-Cresol/Tert-Butanoland p-Cresol/ Piperazinehydrogen Bondcomplexesusing Raman Spectroscopy and DFT Calculation
Huang, M.,
Originally presented on: 11/5/2013 15:48:00 - 16:06:00

Determining Stoichiometry in p-cresol/tert-butanoland p-cresol/ piperazineHydrogen BondComplexesUsing Raman Spectroscopy and DFT Calculation

Min Huang,Lin Cao and Jianghua Wu

Department of Chemistry, Tongji University,1239 Siping Rd., Shanghai200092, China

minontongji@tongji.edu.cn

The formation of the hydrogen bond is important in many chemical processes such as holding strands of DNA together and determining the structure of folded proteins. Raman spectroscopy and density functional theory (DFT) calculation were used to study the association complex of O-H…O in p-cresol/tert-butanol and O-H…N inp-cresol/piperazine. Raman spectra at different molar ratio were obtained to study the band peak shift, band width broadening and band intensity nonlinear changescaused by hydrogen bonding.Using$CHCl_3$ and CCl_4respectively as internal references, quantitative measurements on the selected band intensity of p-cresol/tert-butanol/$CCl_4$and p-cresol/piperazine/$CHCl_3$ternaryallowed the determination of the stoichiometry. The experimental evidences of the 1:1 and 2:1 adductsare presented for the first time using Raman spectroscopy.

(335m) Development of Inorganic Membranes in China - Present and Future
Jin, W., Nanjing University of Technology
Originally presented on: 11/5/2013 16:06:00 - 16:24:00

Development of Inorganic Membranes in China - Present and Future

Wanqin Jin and Nanping Xu

State Key Lab of Materials-Oriented Chemical Engineering,

Nanjing University of Technology, Nanjing 210009, China.

E-mail: wqjin@njut.edu.cn

Inorganic membranes are known for excellent physical and chemical properties, such as thermal stability and chemical durability, which endow them with significant advantages for use in separations that are difficult to achieve by conventional organic membranes, i.e. high-temperature separations and separations of aggressive media. Nowadays inorganic membranes have been widely applied in the fields of chemical, petrochemical, metallurgical, bioengineering, environment engineering, food and drug industry.

Inorganic membranes were first developed in the 2nd World War as gas diffusion barriers for the enrichment of uranium. It is around 1980 that inorganic membranes were used for the separation in liquid media by many countries including the People's Republic of China. Nowadays, with facing immense sustainability challenges in terms of energy, resources and environment, China has made a considerable progress in the inorganic membrane technology. In this talk, a short summary of the development and application of inorganic membranes in China will be presented. These inorganic membranes are applied in various membrane processes, including microfiltration, ultrafiltration, pervaporation, gas separation and membrane reactor. In particular, a number of representative developed membrane materials such as porous ceramic membranes, zeolite membranes, mixed ionic-electronic conducting membranes, metal membranes etc., will be presented here. Furthermore, some typical industrial applications of current membrane processes in China will be illustrated in this presentation, such as environmental protection processes, food and beverage industries, biotechnology and pharmaceutical industries, as well as chemical and petrochemical fields. Finally, future trends of the inorganic membranes of China will be prospected.

(335n) Diversity of Glycyrrhizin Biotransformation By β-Glucoronidases From Fungi

Li, C., Beijing Institute of Technology

Originally presented on: 11/5/2013 16:24:00 - 16:42:00

Diversity of Glycyrrhizin Biotransformation by β-Glucoronidases from Fungi

Chun Li

School of Life Science, Beijing Institute of Technology, Beijing, 100081, P.R. China

Glycyrrhizin (GL), a triterpenoid saponin glycoside and the main constituent of licorice extract (Glycyrrhiza glabra), is a kind of natural edulcorant and sweeter as well as an important ingredient of traditional Chinese medicine. Generally GL can be hydrolyzed into β-D-mono-glucuronide- glycyrrhizin (GAMG) and glycyrrhetic acid, which have higher merits than GL due to its physiological and commercial values. In our previous work, the three β-glucuronidases from *Penicillium purpurogenum* Li-3, *Aspergillus terreus* Li-20 and *Aspergillus ustus* Li-62 could transform glycyrrhizin into GAMG, GAMG and GA, and GA, respectively. The three genes were cloned and also named *p*gus, *at*gus and *au*gus, and not only their alignment results showed significant differences in their N-terminal and C-terminal, but the properties of these enzymes also showed obvious differences.

Wherein a novel β-glucuronidase from *Penicillium purpurogenum* (PGUS) could specially hydrolyze the distal β-D-glucuronic acid residue from glycyrrhizin into GAMG directly, but PGUS-E which *pgus* gene was expressed by *E.coli* BL21(DE3), PGUS-P expressed by *Pichia pastoris* and PGUS-C expressed by CHO cells could sequentially catalyze to release non-reducing terminal β-D-glucuronic acid residues from glycyrrhizin or GAMG into GA. So there was probably predicted a few difference of structures between PGUS and the others. Here we firstly reported the crystal structure of PGUS-E from fungus *Penicillium purpurogenum*. And the crystal structure of PGUS-E has been solved to 3.1 Å in space group c2221. The model of PGUS-E from basically consisted of two dimer in the asymmetric unit which was composed of three domains including the $(a/b)_8$ TIM-barrel motif, a sandwich domain and a typical carbohydrate recognition domain. Furthermore we confirmed that E414 and E505 were the key residues essentially affecting the enzyme activity by site-directed mutagenesis and enzymatic activity assays. These findings also could provide a solid foundation for developing substrate (aglycone) specificity of β-glucuronidases from glycoside hydrolase Family 2.

In addition, the strategies of GAMG production with high efficiency by different types of PGUS were systematically investigated and usefully developed in an ionic liquid/buffer biphasic system and in scale-up fermentation.

(335o) Chemical-Looping Cumbustion of Methane Using Fe2O3-Based Oxygen Carriers With a Thermal Storage Function
Li, K.,
Originally presented on: 11/5/2013 16:42:00 - 17:00:00

Chemical-looping cumbustion of methane using Fe_2O_3-based oxygen carriers with a thermal storage function

Li Kongzhai *, Wang Hua, Wei Yonggang, Zhu Xing, Liu Zisong

(Faculty of Metallurgy and Energy Engineering, Kunming University of Science and Technology, Kunming 650093, China)

Abstract: The chemical-loopingcombustion is a new generation of combustion technology, which owns the advantages ofe high efficiency and low energy consumption for CO_2 capture. Chemical-looping combustion was perfprmed in two reactors: fuel reactor and air reactor. The heat exchange between the two reactors to provide sufficient heat for the endothermic reaction strongly affects the combustio effifiency of the whole process. Thermal energy storage technology is very important to improve energy efficiency and move the gap between energy supply and request.In this paper, thephase change thermal storage technology was used in chemical-looping combustion process throughbuilding a thermal storage functional oxygen carrier, whichcan make the temperature of the oxygen carriers in the fixed bed more stable,and promote the heat exchange bwtween the two reacters.

A series of iron oxide-based oxygen carrierswere prepared by different methods, and the oxygen carriers were characterized by XRD, BET, H_2-TPR, and O_2-TPD techniques. The reactions between oxygen carriers and methane/air were also investigated in detail. The results show that the Fe_2O_3/Al_2O_3 oxygen carrier with a Fe_2O_3 loading of 60wt% prepared bycoprecipitation method revealed the best activity and redox stability for methane combustion. No obvious decline in the conversion of methane and the formation of CO_2 was observed over this oxygen carrier after the redox cycling in alternant methane/air atmosphere for 30 times.Acore-shell heat storage material (Al/AlN) was prepared via the gas-solid reaction between aluminum powder and nitrogen, which were also characterized by TG, DSC, EDS and SEM. DSC results showed that adding of a lithium saltsand increasingthe reaction time could improve the formation of shell AlN. The Al/AlN material using Li_2CO_3 as an addtive and reaction time of 2 h showed a good high-temperature thermalstorageperformance and high anti-oxidation ability. The thermal storage capacity of this oxygen cattier is as high as 225J/g.

It also showed that the addtion of Al/AlN thermal storage material into Fe_2O_3/Al_2O_3oxygen carriers can significantly improve the seletiviry of CO_2in the combustion process of methane. With the Al/AlN content of 80wt%, the heat released in the oxidation step could be totally stored in the Fe_2O_3/Al_2O_3-Al/AlN oxygen carrier. This indiacates that the utilization of thermal storage materials in the chemical-looping combustion process to improve the heat exchange of oxygen carriers between the oxdation and the reduction steps is feasible.

(335p) Green Technical Routes and Clean Productions for Diene-Based High Performance Elatomers
Pan, Q.,
Originally presented on: 11/5/2013 17:00:00 - 17:18:00

Green technical routes and clean productions for diene-based high performance elatomers

Qinmin Pan[1], Zhijun Zhou[2], Garry L. Rempel[3]

[1]Green Polymer and Catalysis Laboratory, Soochow University, Suzhou 215123, PR China; [2]Department of Chemical and biological Engineering, Zhejiang University, Hangzhou 310027, PR China; [3]Department of Chemical Engineering, University of Waterloo, Waterloo, Ontario, N2L 3G1, Canada

This work is focused on investigating a green polymerization and hydrogenation route for the synthesis of hydrogenation nitrile butadiene rubber (HNBR), which is a very important high performance elastomer and widely used in automobile and petrochemical industry.

The current commercial process of HNBR starts with emulsion polymerization followed by solution hydrogenation. Thus complicated pretreatment and post-treatment are involved, and the process has not been optimized in terms of the energy cost and material cost, and the environmental impact is not yet minimized. In our current work, we successfully realized the hydrogenation of NBR directly in the latex form, and thus the use of the organic solvents that are required in the traditional process are eliminated, and the polymerization and the hydrogenation can be integrated into a single process. Furthermore, we developed a membrane separation process to efficiently separate surfactants and purify the waste water so that they can be recycled and reused at the start points. A hydrogenation degree higher than 95% without any crosslinking, which is required for the product to reach the qualification index as high performance elastomer, is relized. 3Rs (reduce, reuse and recycle) are completely achieved without any waste liquid/water to be discharged, i.e. zero discharge.

(335q) Extraction of Fir Wood Oil From Fir Wood Waste and Synthesis of Cedar Series Perfumes

Zhu, K.,

Originally presented on: 11/5/2013 17:18:00 - 17:36:00

Extraction of fir wood oil from fir wood waste and synthesis of Cedar series perfumes

Zhukai:zhukai53@163.com

Nanjing Forestry University

Cedar wood oil is one of the top ten essential oils In the world, Cedar wood oil can be btained from fir wood wast, high added value Cedar series perfumes are btained from Cedar wood oil by Chemical processing,and their aroma is very unique, and cannot be replaced. At present,the Cedar wood resources are dwindling, a serious influence had taken place in the flavor and fragrance industry. Study found that essential oil is very rich in fir wood waste, Cedar wood oil can be replace completely with the fir oil to used in perfume industry according to the aroma and chemical composition. The fir oil was extracted from fir root and its processing residu by the steam distilling and retorting methods.The retorting essential oil,wood vinegar, wood tar, charcoal and combustible gases, were obtained from fir waste by retorting. the main perfumes such as Cedrol, Cedrene, Thujopsene, etc.were made from the oil by refining and isolating,next to Chemical processing, dozens of Cedar series perfumes with high added value, such as acetyl cedrene, cedrol methyl ether, etc. were synthesized.

(335r) High-Efficiency Synthesis of Cyclic Carbonates From Epoxides and CO2 Over Immoblized Hydroxyl Ionic Liquid Catalyst

Yin, S.,

Originally presented on: 11/5/2013 17:36:00 - 17:54:00

High-Efficiency Synthesis of Cyclic Carbonates from Epoxides and CO_2 over Immoblized Hydroxyl Ionic Liquid Catalyst

Wei-Li Dai[1], Chak-Tong Au[2], Shuang-Feng Yin[1,*]

([1] College of Chemistry and Chemical Engineering, Hunan University, Changsha 410082, China; [2]Department of Chemistry, Hong Kong Baptist University, Kowloon Tong, Hong Kong, China)

*Corresponding author. Shuang-Feng Yin, Tel/ Fax: 86-731-88821171, E-mail: sf_yin@hnu.edu.cn(S.F. Yin).

The fixation of CO_2 into valuable compounds has attracted much attention. Of the most interesting is to synthesize five-membered cyclic carbonates by the coupling of CO_2 with terminal epoxide. Many catalysts have been reported for the coupling reaction, including quaternary ammonium salts, metal oxide, molecular sieve, and so on. Due to the disadvantages related to product separation using homogeneous catalysts, heterogeneous catalysts have attracted much attention. In the past ten years, ionic liquid attracted much interest in their uses as catalysts for synthesizing cyclic carbonates. As a strategy for the fabrication of stable and reusable heterogeneous catalysts, ionic liquids were immobilized on solid materials such as MgO, SiO_2, and MCM-41. However, catalyst reusability is far from satisfaction.

Herein, we report the fabrication and characterization of hydroxyl ionic liquid grafted onto cross-linked divinylbenzene polymer (PDVB-HEIMBr). It was evaluated as a catalyst for the synthesis of cyclic carbonates from CO_2 and epoxides without the use of any co-catalyst and organic solvent. The catalyst shows good performance across a wide range of epoxides, giving almost quantitative yield of carbonates. The effects of reaction temperature, time and initial CO_2pressure on product yield were investigated. It is suggested that the synergetic effect between the bromide ions and the hydroxyl groups facilitates the coupling reaction. Furthermore, the PDVB-HEIMBr catalyst shows excellent stability and reusability. From the viewpoint of industrial application, the catalystis very attractive because of its simplicity, activity, stability, and reusability.

Scheme 1

(342a) Immobilization of Recombinant Antibody Fragment Utilizing Material-Binding Peptides for Sensitive Immunoassay and Immuno-Sensors

Kumada, Y., Kyoto Institute of Technology
Kishimoto, M., Kyoto Institute of Technology
Originally presented on: 11/5/2013 15:20:00 - 15:40:00

Recombinant antibody fragments such as Fab, scFv and VHH are recently considered as next generation of diagnostic reagents because they can be immobilized with high density on the surfaces of materials, and consequently, antigen molecules would be efficiently captured by these ligand antibodies. However, such antibody fragments have not been utilized as a ligand antibody in enzyme immunoassay (EIA) because of difficulty in site-specific immobilization of antibody fragments with high remaining antigen-binding activities due to conformational change of antigen-binding domains after immobilization. In order to immobilize recombinant antibody fragments with high remaining activity, site-specific attachment of antibody fragments through the material-binding peptides is necessary. In this study, we originally identified material-binding peptides that can strongly interact with the surface of polystyrene (PS), polymethyl methacrylate (PMMA), and silicon nitride (SiN). By introducing these peptides to the C-terminus of recombinant antibody fragments, it was possible to increase both density as well as remaining activity of the fragments after passive adsorption to the solid materials. We also investigated efficient refolding method for peptide-tagged antibody fragments and thus, all of peptide-fused antibody fragments could be refolded with over 90 % of recovery, followed by site-specific immobilization. Our developed methods for refolding and immobilization of material-binding peptide-fused antibody fragments are considerably useful for preparation of high-performance of antibody-immobilized solid-support for immunoassay and immune-sensors.

(342b) Biomagnetic Nanomaterials for High Gradient Magnetic Affinity Separation

Lee, G. U., University College Dublin
Fields, C., 1School of Chemistry and Chemical Biology, University College Dublin
O'Mahony, J., UCD
Matias, V., UCD
Li, P., University College Dublin
Smith, L., UCD
Muzard, J.,
Originally presented on: 11/5/2013 15:40:00 - 16:00:00

The growth of biopharmaceutical and biomedical industries has created a demand for new technologies for purification of genetically engineered proteins. High gradient magnetic separation (HGMS) is a technology that has the potential to increase the speed of traditional affinity chromatography due to its use of an expanded bed of superparamagnetic microparticles (SPM). The efficiency of large-scale HGMS is greatly improved if SPMs offering high binding capacity, high magnetic susceptibility, and high magnetization are employed. In this communication, we will describe strategies for synthesizing SPMs that are composed of biomolecules and viruses assembled on a superparamagnetic core. Separation of specific proteins and antibodies from high-protein concentration solutions was performed in a rapid and single step. The dense SPM core of these particles makes them highly responsive to magnetic fields and the biomolecular coating produces a high binding capacity. These new biomaterials and the described technology appear to be well suited to large-scale HGMS separation and promise to be cost effective.

(342c) Theoretical Investigations of Multiple Weak Interactions in Multimodal Chromatography Using Molecular Dynamics Simulations and Protein Surface Characterization

Parimal, S., Rensselaer Polytechnic Institute
Woo, J. A., Rensselaer Polytechnic Institute
Srinivasan, K., Rensselaer Polytechnic Institute
Holstein, M. A., Rensselaer Polytechnic Institute
Garde, S., Rensselaer Polytechnic Institute
Cramer, S. M., Rensselaer Polytechnic Institute
Originally presented on: 11/5/2013 16:00:00 - 16:20:00

Multimodal (MM) chromatography has emerged as an extremely promising form of chromatography. Recent studies published by our group and others have shown that an appropriate choice of ligand chemistry and fluid phase modifiers can modulate the different modes of interactions present in MM systems to achieve enhanced selectivities as compared to traditional single-mode chromatography. However, exciting as the experimental results might be, there is still a lack of mechanistic details in these MM systems which is essential to optimize desired separations. The goal of the work presented here is to provide fundamental understanding of the nature of water-mediated protein-ligand interactions at a molecular level which can provide significant insight into the design of MM ligands, the roles of synergy and the modulation of selectivity using fluid phase modifiers (FPMs).

A combination of theoretical techniques has been used to understand MM interactions. Molecular Dynamics (MD) simulations have been employed to gain molecular insights into how multiple weak interactions in MM systems work together to create binding selectivity. All-atom explicit MD simulations have been performed with two different MM ligands interacting with proteins and the results are analyzed to understand binding preferences of these MM ligands in terms of the interaction chemistry and their conformational flexibility. Simulations have also been performed with different fluid phase modifiers to understand how co-solutes affect protein-ligand binding and to shed light on the different mechanisms which govern their behavior.

Protein surface characterization techniques have also been employed as a fast, coarse-grained approach to map proteins based on their potential to interact with different modes. Protein-ligand binding in MM systems involves a combination of electrostatic and hydrophobic interactions. To accurately measure the individual contributions of both these modes, and their possible synergistic effects to the overall binding behavior requires the characterization of the molecules involved in binding in terms of their electrostatic and hydrophobic potential. While electrostatic interactions between two molecules can be easily calculated, hydrophobicity has been difficult to understand. Hence, hydrophobic characterization of proteins has been carried out here using recently developed methods which incorporate the "context-dependence" of hydrophobicity. The results from these calculations, together with the molecular understanding gained from

MD simulations, have been used to explain chromatographic retention behavior of proteins in MM systems on different resins and under different fluid phase conditions.

The work presented here demonstrates the use of theoretical techniques, especially MD simulations, to probe atomistic details of protein-ligand binding which can shed light on and indeed predict protein retention behavior in chromatographic systems. Finally, the knowledge base created using these simulations can be used to select appropriate combinations of MM ligands and modifiers to achieve unique selectivities for challenging protein separations.

(342d) Development of Adsorptive Membranes With Glycoligands for Targeted Lectin Binding

Chenette, H. C. S., Clemson University
Husson, S. M., Clemson University
Originally presented on: 11/5/2013 16:20:00 - 16:40:00

This presentation will describe our latest efforts to develop adsorptive membranes for use as protein purification media. The design approach used in these studies is to graft polymeric 'tentacles' from macroporous regenerated cellulose membranes by atom transfer radical polymerization (ATRP). Using one or more examples, we will illustrate how this design approach can be used to prepare high-productivity membrane adsorbers. In one case, we will discuss affinity membrane adsorbers that incorporate glucose-containing glycoligands. Studies have shown glycoligands selectively bind specific lectins [Yang et al., *Appl. Mater. Interfaces*, **2010**, *2*] mimicking the gluco-receptors found on cell surfaces that recognize and bind bacteria. The literature also reports advances in this field using model substrates to elucidate optimal polymerization conditions, and kinetic studies relating polymerization time to the degree of polymerization [Yang and Ulbricht, *Macromolecules*, **2011**, *44*]. The present study integrates these glycoligand adsorbers into a membrane system that is capable of scale-up to a column for chromatographic separation. A glucose-based ligand was grafted from regenerated cellulose membrane surfaces using ATRP. The model lectin, Concanavalin A, was used to evaluate membrane performance in bind-and- elute purification, using a low molecular weight sugar for elution.

(342e) Overload and Elute Chromatography for Enhanced MAb Purification

Nadarajah, D., Genentech, Inc.
Hohwald, S., Genentech, Inc.
Banerjee, D., Genentech, Inc.
Shomglin, K., Genentech, Inc.
St. John, R., Genentech, Inc.
Mehta, A., Genentech, Inc.
Originally presented on: 11/5/2013 16:40:00 - 17:00:00

Bio-pharmaceutical process development has increasingly been relying on platform processes that reduce development timelines and time to clinic. However, certain MAbs may not fit into the platform process thus requiring further development. This talk will focus on a set of MAbs that bind to the anion exchange (AEX) resin under flow-through conditions. To enable plant fit, improve yield and achieve adequate impurity clearance, a novel mode of chromatography was developed with a multimodal resin. The resin was operated in an overload mode with a step elution enabling significantly higher load density while obtaining the desired impurity clearance. Data on clearance of model viruses, host cell proteins and aggregates will be presented. High throughput screening was employed to rapidly identify process parameters for overload and elute chromatography. Data generated to develop a mechanistic understanding of product and impurity binding to the resin under the overload and elute mode of chromatography will also be presented.

(342f) Development of Production Method for Zeaxanthin By Fermentation Using Paracoccus Carotinifaciens

Nagai, H., Ajinomoto Co., Inc.
Originally presented on: 11/5/2013 17:00:00 - 17:20:00

Zeaxanthin is a yellow carotenoid contained in Corn, Paprika, Spinachi, Chinese matrimony vine, hen egg and so on. In human body, this carotenoid exists especially in the center of macular region on retina.

Age-related macular degeneration (AMD) causes the reduction of the content of Zeaxanthin on the macular of patients. Also, many researchers reported that most of these kinds of eye diseases are caused by oxidative stress. Since Zeaxanthin is a strong antioxidant, we can imagine that this carotenoid has a certain kind of prophylactic effect on human eyes.

In many situations, supplements are generally used for the uptake of Zeaxanthin into human blood, because the general foods contain the less content of Zeaxanthin. However, this process generally shows very low absorption efficiency while the concentration of zeaxanthin in the blood can be kept at high level (0.48□ 1.09μM). Since supplements are not inexpensive, this can be a barrier for the people who want to prevent falling to those eye diseases.

□However, it is possible to consider hen egg as a supply source of Zeaxanthin. In this case, Zeaxanthin concentration in the blood is very low(0.06□0.15μM), while absorption efficiency is very high. If we can increase the content of Zeaxanthin in hen egg, hen egg can be a candidate of the most efficient for Zeaxanthin supply for human.

Already some developers have used spirina or Chinese matrimony vine to produce Zeaxanthin enriched egg. However, the maximum its concentration of Zeaxanthin was less than 3.35mg/100g. In order to solve this provlem, we developed the dried Paracoccus cell containing high content of Zeaxanthin (Panaferd-ZX®). Addition of Panaferd-ZX into feed for hen shows higher content of Zeaxanthin in hen egg (3.5mg□ 10mg/100g).

In order to develop the production method, we selected some strains producing Zeaxanthin from mutant Astaxanthin production strains. Furthermore, we have developed fermentation process using these strains.

For the scale-up of fermentation process, we have faced to insufficient cell growth. From laboratory scale analysis, we have found that these strains showed insufficient growth under the pressure ranged from 0.02MPa to 0.08MPa. Since the large scale fermenter has the pressure up to 0.1MPa by the level of fermentation broth, it is possible that this is the reason of unsuccessful fermentation. To solve this problem, we examined the additives for fermentation media. Finally, we have found that citric acid can promote the cell growth under the pressure. This enabled us successful scale-up to 5kL scale fermenter and trial sales.

(342g) Molecular and Pore Diffusion Coefficients of DNAs and Modified Proteins for Process Chromatography

Ohshima, S., Yamaguchi University
Yoshimoto, N., Yamaguchi University
Yamamoto, S., Yamaguchi University
Originally presented on: 11/5/2013 17:20:00 - 17:40:00

As the performance of chromatography is governed by pore diffusion, it is important to know the pore diffusion coefficient especially for separation of large bioproducts such as proteins and DNAs.

However, compared with the diffusion coefficient of globular proteins (molecular weight 10,000- 100,000), the diffusion coefficients of DNAs and modified proteins have not been measured extensively. Consequently, it is difficult to predict the diffusion coefficient in the solution (molecular diffusion coefficient, D_{m}), which is needed for understanding the pore diffusion coefficient. D_{s}.

In this study, D_{m} was determined by the Taylor method, which uses a pulse response curve from a narrow tube at laminar conditions. Single strand and Double stranded DNAs were used as the sample. D_{s} was determined also by using a pulse response experiment with agarose-based ion-exchange chromatography particles, Sepharose HP under non-binding conditions.

D_{m} of DNA was correlated with the molecular weight M_{W} as follows.

Single strand DNA : $D_m = 1.94\mathrm{x}10^{-8} M_w^{-0.57}$

Double strand DNA : $D_{\mathrm{m}} = 1.33\mathrm{x}10^{-8} M_w^{-0.49}$

Compared with the established correlation for proteins (D_{m} is proportional to $M_w^{-0.33}$), D_{m} changes much more strongly with M_{W}. This is because of the shape of DNA molecules.

D_{s} of single stranded DNA was proportional to $M_w^{-0.99}$, which indicates that diffusion of single strand is highly hindered in the pore. This also suggests that separation of DNA molecules with conventional porous particles is difficult and the use of convection-aided media such as monoliths is preferred.

(374a) Simulation Tools for Nanoparticle-Based Composite Processing and Property Prediction

Bolintineanu, D. S., University of Minnesota
Lechman, J. B., Sandia National Laboratories
Schunk, P. R., Sandia National Laboratories
Originally presented on: 11/5/2013 15:15:00 - 15:40:00

We present a set of tools for the modeling of various aspects of solution-based nanomanufacturing processes. First, we discuss several discrete element modeling (DEM) techniques for the simulation of nanoparticle/colloidal suspensions, including efficient algorithms for the inclusion of hydrodynamic interactions, capillary forces, long-range colloidal interparticle forces, granular contact forces, solvent evaporation and highly aspherical particles. We apply these techniques to the simulation of a battery electrode manufacturing process, and show that the predicted microstructures provide relevant insights for the processing parameters. In the second half of the talk, we discuss methods for the prediction of the bulk properties of disordered, particle-based composites. We present several approaches that rely on a finite element model (FEM) representation of a sample of the material, which can be extracted from a DEM simulation or taken from a tomographic image of the real composite. Overall, this work presents an extensive, multiscale simulation toolset for connecting particle-based nanomanufacturing processing conditions to resulting microstructures and ultimately to material properties.

(374b) Tin Oxide Nanowires and Their Hybrid Architectures for Kinetically Fast Redox Couples in Dye-Sensitized Solar Cells

Vendra, V. K., University of Louisville
Nguyen, T., University of Houston
Druffel, T., Conn Center for Renewable Energy Research
Sunkara, M., University of Louisville
Amos, D. A., University of Louisville
Originally presented on: 11/5/2013 15:40:00 - 16:05:00

The redox electrolyte is an important component of the dye-sensitized solar cells. Iodide/tri-iodide electrolyte used in high efficiency cell is not suitable for practical application due to its corrosive nature, toxicity, competition between dye and tri-iodide for light aborption. One of the major drawback of using alternate redox couples is their fast recombination kinetics and mass transport limitations.

In this study, we show that tin oxide nanowires show a substantial improvement in the photovoltaic performance when compared to titania nanoparticles, when electrolytes having fast recombination kinetics are employed in dye-sensitized solar cells (DSCs). State of the art methods in literature employ titania nanoparticles use passivating layers of Al_2O_3 or poly(methyl siloxane) to lower the electron recombination rate. There have been reports investigating the performance of nanowire in conjunction with alternate redox couples. The single crystalline nature allows for fast electron transport in tin oxide nanowires. Further, prior work from our group has also shown that tin oxide nanowires exhibit two orders of magnitude higher electron lifetimes when compared to titania nanoparticles when iodide/tri-iodide is used as an electrolyte. In this study we employed tin oxide nanowires as photoanodes for DSCs and investigated their performance when kinetically fast electrolytes such as ferrocene/ferrocenium,tempo/tempo,ferricyanide /ferrocyanide are used as redox electrolytes.

Surface modifications such as $TiCl_4$ treatment and atomic layer deposition of titania and alumina were performed to further lower electron recombination in tin oxide nanowires. Fundamental properties such as electron transport time scales and electron recombination time scales were determined for tin oxide nanowires and hybrid architectures where the surface of tin oxide nanowires is coated with titania and alumina.

(374c) Multiscale Sustainability Assessment in Nanocoating Material Design and Manufacturing

Song, H., Wayne State University
Uttarwar, R., Wayne State University
Huang, Y., Wayne State University
Originally presented on: 11/5/2013 16:05:00 - 16:30:00

Nanopaint has drawn great attention in the coatings industry, as it may offer surface coatings with significantly improved performance or new functionalities and properties, such as surface texture alteration control, self-healing, toxic chemical/acid/corrosive agent repelling, glass opaqueness control, building exterior's infrared reflecting/absorbing properties control, and even pathogen growth control. Nevertheless, it is under scrutiny for the environmental and health risks which may pose in nanopaint manufacturing and nanocoating realization. Therefore, it is crucial to analyze the sustainability performance of nanocoating materials and prepare a roadmap for their successful development and application in the future.

In this paper, we introduce a systematic sustainability assessment methodology by resorting to multiscale science and sustainability science. Different from known methodologies focusing on the assessment of manufacturing process and manufactured product, this methodology initiates the assessment starting at the nanomaterial formulation design stage. A variety of material design parameters, such as those about polymer system, nanomaterial type, shape, size, and orientation, solvent and additives, are analyzed using a number of sustainability indicators, which will be incorporated into the analysis of coating manufacturing stages. Multiscale models are developed to predict nanoparticle emission severity in nanopaint spray and curing, nanocoating properties, product realization process performance on quality assurance, waste reduction, and operational safety. The efficacy of the multiscale sustainability assessment methodology will be demonstrated by the case studies on nanocoatings with super-scratch resistance performance and barrier property. The introduced methodology is generic and can be applied to study the sustainability performance of a wide range of nanocoating materials in their early design stage.

(374d) Continuous Nanoparticle Sizing and Characterization Via Microfluidic Interfacial Fluorescent Complexation

Meng, F., Texas A&M University
Ugaz, V. M., Texas A&M University
Originally presented on: 11/5/2013 16:30:00 - 16:55:00

A variety of characterization tools are available to sensitively analyze collected nanoparticles (e.g., condensation particle counters, differential mobility analyzers, scanning mobility particle sizers (SMPS), ion chromatography, mass spectrometry, and electron microscopy). But these methods often require a dedicated laboratory infrastructure to operate optimally, and are therefore not straightforward to adapt for online use. We have developed a new microfluidic-based approach that overcomes many of these limitations by harnessing the inherently steep chemical gradient established at the interface between co-flowing streams containing a nanoparticle-laden suspension and a fluorescent dye. This sharp mismatch acts to localize adsorptive dye-nanoparticle complexation interactions within a narrow interfacial zone, instantaneously producing an intense and easily detectable fluorescence signature. These interactions are inherently dominated by phenomena occurring at the nanoparticle surface, introducing the exciting possibility to extract information about particle size and morphology from the fluorescence profile. We demonstrate this capability experimentally using ZnO and TiO_2 nanoparticles, and introduce a reaction model that enables interfacial fluorescence to be predicted so that kinetic parameters associated with the underlying surface complexation can be quantified. Our approach offers potential for continuous operation that is particularly attractive in manufacturing settings where robust on-line characterization tools are needed.

(374e) Safer Formulation Concept for Flame-Generated Engineered Nanomaterials

Sotiriou, G. A., ETH Zurich
Gass, S.,
Cohen, J., Harvard University, School of Public Health
Pyrgiotakis, G., University of Florida
Pratsinis, S. E., ETH Zurich
Demokritou, P., Harvard School of Public Health
Originally presented on: 11/5/2013 16:55:00 - 17:20:00

The likely success or failure of the nanotechnology industry depends on the environmental health and safety of engineered nanomaterials (ENMs). While efforts toward engineering safer ENMs are sparse, such efforts are considered crucial to the sustainability of the nanotech industry. A promising approach in this regard is to coat potentially toxic nanomaterials with a biologically inert layer of amorphous SiO_2. Core-shell particles exhibit the surface properties of their amorphous SiO_2 shell while maintaining specific functional properties of their core material. A major challenge in the development of functional core-shell particles is the design of scalable high-yield processes that can meet large-scale industrial demand. Here, we present a safer formulation concept for flame-generated ENMs based on a one-step, in flight SiO_2 encapsulation process, which was recently introduced by the authors as a means for a scalable manufacturing of SiO_2 coated ENMs. Firstly, the versatility of the SiO_2-coating process is demonstrated by applying it to four ENMs (CeO_2, ZnO, Fe_2O_3, Ag) marked by their prevalence in consumer products as well as their range in toxicity. The ENM-dependent coating fundamentals are assessed and process parameters are optimized for each ENM investigated. The effects of the SiO_2-coating on core material structure, composition and morphology, as well as the coating efficiency on each nanostructured material, are evaluated using state-of-the-art analytical methods (XRD, N_2 adsorption, TEM, XPS, isopropanol chemisorption). Finally, the biological interactions of SiO_2-coated vs. uncoated ENMs are evaluated using cellular and animal bioassays, providing valuable evidence for reduced toxicity for the SiO_2-coated ENMs. Results indicate that the proposed 'safer by design' concept bears great promise for scaled-up application in industry in order to reduce the toxicological profile of ENMs for certain applications.

References

[1] Sotiriou G.A., Sannomiya T., Teleki A., Krumeich F., Vörös J. and Pratsinis S.E. Non-toxic dry-coated nanosilver for plasmonic biosensors. *Adv. Funct. Mater.* **20**, 4250-4257, (2010).

[2] Sotiriou G.A., Hirt A.M., Lozach P.Y., Teleki A., Krumeich,F. and Pratsinis S.E. Hybrid, silica-coated,

Janus-like plasmonic-magnetic nanoparticles. *Chem. Mater.* **23**, 1985-1992, (2011).

[3] Gass S., Cohen J.M., Pyrgiotakis G., Sotiriou G.A., Pratsinis S.E. and Demokritou P. Safer Formulation Concept for Flame-Generated Engineered Nanomaterials. *ACS Sustainable Chem. Eng.* **in press**, DOI: 10.1021/sc300152f, (2013).

[4] Demokritou P., Gass S., Pyrgiotakis G., Cohen J.M., Goldsmith W., McKinney W., Frazer D, Ma J, Schwegler-Berry D, Brain J, Castranova V., An in vivo and in vitro toxicological characterisation of realistic nanoscale CeO_2 inhalation exposures *Nanotoxicology* **in press**, DOI: 10.3109/17435390.2012.739665 (2013).

(374f) Green Pathways for Development of Nanostructured Aerogel Photocatalysts Effective At Both UV and Visible Range

Li, H., University of South Florida
Sunol, S. G., University of South Florida
Sunol, A. K., University of South Florida
Originally presented on: 11/5/2013 17:20:00 - 17:45:00

Nano-structured highly porous $TiO_2/WO_3/Fe^{3+}$ aerogel photocatalyst are prepared, characterized, and tested for model photo-catalytic reactions. The catalyst structure is tailored to capture environmental pollutants and destruct them under both u-v and visible light through oxidation.

A novel and green method is utilized to prepare the unique surfactant templated aerogel photo-catalyst that has highly porous nanostructure with high surface area and tailored pore size distribution. Sol-gel process is combined with supercritical extraction, deposition and drying. Supercritical drying with heat treatment results in titanium dioxide anatase crystal form. Templates used further enable retention and tuning of the nano-pore structure and surface properties.

The catalyst was characterized using AFM, SEM, FIB, XRD, XPS, BET and porosimetry prior to using in reactions. The band gap of the catalyst particles were determined using diffuse reflectance.

Resulting aerogel $TiO_2/WO_3/Fe^{3+}$ has similar photo-catalytic capability compared to highly optimized commercial Degussa P25 under UV exposure and offer much superior photo-catalytic capability under visible light exposure. The model reaction utilized employed Methylene Blue photo oxidation under Visible and UV light.

Engineering Cell Niches in a Couple of Clicks

Anseth, K. S., University of Colorado-Boulder

Originally presented on: 11/5/2013 18:00:00 - 18:25:00

Methods for culturing mammalian cells in a biologically relevant context are increasingly needed to study cell and tissue physiology, expand and differentiate progenitor cells, and to grow replacement tissues for regenerative medicine. Two-dimensional culture has been the paradigm for *in vitro* cell culture; however, evidence and intuition suggest that cells behave differently when they are isolated from the complex architecture of their native tissues and constrained to petri dishes or material surfaces with unnaturally high stiffness, polarity, and surface to volume ratio. As a result, biologists are often faced with the need for a more physiologically relevant 3D culture environment, and many researchers are realizing the advantages of hydrogels as a means of creating custom 3D microenvironments with highly controlled chemical, biological and physical cues. Further, the native ECM is far from static, so ECM mimics must also be dynamic to direct complex cellular behavior. In general, there is an un-met need for materials that allow user-defined control over the spatio-temporal presentation of important signals, such as integrin-binding ligands, growth factor release, and biomechanical signals. Developing such hydrogel mimics of the ECM for 3D cell culture is an archetypal engineering problem, requiring control of numerous properties on multiple time and length scales important for cellular functions. New materials systems have the potential to significantly improve our understanding of how cells receive information from their microenvironment and the role that these dynamic processes may play in controlling the stem cell niche to cancer metastasis. This talk will illustrate our recent efforts to advance hydrogel chemistries for 3D cell culture and dynamically control biochemical and biophysical properties through orthogonal, photochemical click reaction mechanisms.

(459a) Theory and Computation in Modern Chemical Engineering. A Thermodynamicist's Perspective

Debenedetti, P. G., Princeton University
Originally presented on: 11/6/2013 11:15:00 - 11:40:00

Chemical engineers have been at the forefront of the development of powerful algorithms that enable the molecular-based analysis of complex systems, such as therapeutic protein formulations and fuel cell membranes. Building upon a solid foundation in thermodynamics and statistical mechanics, these methods allow detailed microscopic scrutiny of a wide range of problems of interest in modern chemical engineering, thereby providing fundamental understanding and facilitating rational design. I will illustrate the insights that can be gained from theory and computation by drawing on examples from my own research in areas including chiral drug resolution and the solubility of trace pollutants in water.

(461a) US Department Of Energy Critical Materials Strategy and R&D

McKittrick, M.,

Originally presented on: 11/6/2013 12:30:00 - 12:55:00

In 2010, the US Department of Energy (DOE) released its first Critical Material Strategy – which examined the role of rare earth elements and other materials in the clean energy economy. The Strategy, which was also updated in 2011, highlighted five elements as being "critical" to the deployment of clean energy technology – dysprosium, europium, neodymium, terbium and yttrium. Building off this work and leveraging R&D investments in the area, in 2013 DOE established the Energy Innovation Hub for Critical Materials. This Hub, led by Ames Laboratory, will conduct R&D across the entire lifecycle including materials discovery and design, feedstock supply, processing, manufacture, use, recycling, and re-use. A summary of the DOE Critical Materials Strategy and the Critical Materials Hub will be presented.

(461b) Research Gaps and Needs: Critical Materials Separations R&D Workshop

Schuster, D., Institute for Sustainability
Hunt, C. T., Emerging Technologies, Rohm and Haas Company
Chin, K., AIChE
Diallo, M., Caltech
Originally presented on: 11/6/2013 12:55:00 - 13:20:00

Report from recent workshop on research 'gaps and needs' in separation science and technology for sustainable extraction, recycling, and re-use of critical materials and alternatives will be presented. The enabling role of Separations Science and Engineering is critical in ensuring sustainable supply and utilization of critical materials and the alternatives.

(461c) The Availability Of Indium and Tellurium For Thin-Film Photovoltaic Materials

Eggert, R.,
Lokanc, M.,
Redlinger, M.,
Originally presented on: 11/6/2013 13:20:00 - 13:45:00

Indium and tellurium provide essential properties in emerging thin-film photovoltaic materials. These elements are geochemically scarce. Both are produced today almost exclusively as byproducts of zinc (in the case of indium) and copper (in the case of tellurium) production. If demand for these elements increases significantly, to what degree will new supplies be able to meet this demand? Will increased supplies come from expanded byproduct production at costs similar to recent prices? If not, where might incremental supplies come from and at what costs?

(461d) Rare Earth Concentration, Extraction, Separation and Reduction
Anderson, C.,
Originally presented on: 11/6/2013 13:45:00 - 14:10:00

Rare Earth Concentration, Extraction, Separation and Reduction.

Dr. Corby G. Anderson

Abstract

The key characteristics that deem the Lanthanides to be called Rare Earths focus on their inherent difficulty and challenges in beneficiation, separation and reduction.

This presentation will address the fundamentals behind these processing challenges along with the existing and potential technologies available to remediate them.

(461e) Critical Materials Recycling & Substitutes For Critical Materials

Jones, E., Lawrence Livermore National Laboratory

Originally presented on: 11/6/2013 14:10:00 - 14:35:00

The DOE Critical Materials Institute (CMI) is focused on increasing supply and reducing demand for rare earth elements in clean energy technologies. Approximately 60% of rare earths resourced from mining end up back in landfill or construction materials. This wastage comes from the manufacturing processes or disposals from the consumer reservoir. Thus, finding substitute manufacturing processes and increasing recyclability and reuse can substantially impact the sustainability of rare earth applications. This talk will discuss the R&D approaches that CMI is taking with respect to recycling and substitute technologies, as well as system analytics to improve the lifecycle sustainability of critical materials.

(461f) Critical Materials Recycling For Resource Sustainability

Mishra, B.,

Originally presented on: 11/6/2013 14:35:00 - 15:00:00

Rare earth metals, including yttrium and scandium, are being increasingly used in clean energy technologies, colored phosphors, lasers and high intensity magnets. There are important defense applications such as fighter jet engines, missile guidance systems and space based satellite and communication systems based on these metals. The commitment to clean energy technologies by governments and the projected growth in power and transportation sectors across the globe ensure that the demand for rare earth metals and compounds would continue to escalate. This demand implies that, to ensure unhindered technological innovation, it is essential to possess secure supply chains for rare earth elements. The United States continues to be one of the largest consumers and importer of rare earths and the trend is expected to continue as the demand increases. In order to ensure secure rare earth supply and attenuate supply-demand imbalance post 2014, it is not only necessary to encourage and support exploration of newer reserves, build a rare earth stockpile, but it is also of utmost importance to look at opportunities to recycle and reuse Rare Earth Elements from secondary sources, such as post-consumer and manufacturing process wastes. This research describes the technological developments made to convert these valuable resources into functional manufactured materials for lighting industry, automotive and petroleum refining catalysts, and high density permanent magnets. In addition, production of rhenium from advanced aerospace alloys will be discussed that is recovered for introduction in turbine alloys.

(478a) Enhanced Yield of Bio-Gas Production With Modified Themo-Reactor Configuration and Separation of CO2

Tahir, M. S., University of Gujrat
Saleem, M., Institute of Chemical Engineering and Technology, University of the Punjab, Lahore, Pakistan
Originally presented on: 11/6/2013 12:30:00 - 12:50:00

Pakistan is currently facing severe energy crisis with significant shortfall. A number of short-term technical solutions are available but are not adequate to solve the issue in the longer run. Production of bio-gas with modified thermo-reactor from the animal waste at regional level shows potential of reducing burden on total natural gas resource in Pakistan.

This research study investigates the effects of solar thermo-rector configuration on the overall yield and composition of bio-gas from dung. The designed capacity of the bio-gas plant is 10 m^3. The solar thermo-reactor employs internal heating system with coils using warm water heated from the solar energy. The bio-gas reactor was operated over the range of 25-55 ^{0}C to observe effects on bio-gas yield and composition. It was observed that the increase in temperature within the thermo-reactor resulted in significant increase in yield of bio-gas i.e. thermo-reactor operated at 50-55 ^{0}C (thermophilic). Operation time and plant size of state of the art process can be reduced.

The calorific value of bio-gas obtained at 50-55 ^{0}C was slightly higher than calorific value of bio-gas obtained at mesophilic condition (25-35°C) due to increased production of CH_4. Thermophilic operation of bio-gas plant also enhances the methane content in the bio-gas i.e. 57-60 % vol of methane. Bio-gas produced was purged from bottom and cold water from top a packed column and decreases in CO_2 was also recorded. The results were scaled up from pilot scale capacity to regional level in order to estimate potential of bio-gas production

(478b) Microwave-Assisted Catalytic Glycolysis of Poly(ethylene terephthalate) By Using Ionic Liquids

Liu, Y., Donghua University
Xing, Y., Donghua University

Originally presented on: 11/6/2013 12:50:00 - 13:10:00

Poly(ethylene terephthalate) (PET) was widely used for beverage bottles, electrical/electronic instruments and household wares. As a consequence of dramatically increasing consumption, recycling of post-consumer PET products has become an important environmental opportunity for sustainable usage in society. The presented study is aiming to investigate the use of ionic liquids as catalysts for the glycolysis of PET. A series of novel ionic liquids using 1-butyl-3-methylimidazolium chloride (BMIMCl) and zinc halides (ZH) as precursor with different molar ratios was first synthesized. The glycolysis of PET by microwave-assisted method using the synthesized ionic liquids as catalysts and glycol as solvent was also studied. The influences of the type of zinc halide, the molar ratio of BMIMCl:ZH, the dosage of ionic liquid and solvent, reaction time and reaction time on the yield of recovered terephthalic acid were investigated. The synthesized ionic liquids and the glycolysis product were characterized by X-ray diffractometer, infrared spectrometric analyzer, ^{1}H NMR, differential scanning calorimeter and DSC. The results showed that the synthesized ionic liquids exhibited higher catalytic activity for the glycolysis of PET by microwave-assisted method□The halogen anion has a key influence on the glycolysis of PET. A study on the recycling of the ionic liquid shows that it could be used repeatedly. A mechanism of the degradation of PET was proposed and the kinetics of the reaction was needed further investigated.

(478c) Degradation of Rhodamine-B By Heteropolyoxometalate Ionic Liquid Under Visible Light

Liu, Y., Donghua University
Chen, Y., Donghua University
Xing, Y., Donghua University
Originally presented on: 11/6/2013 13:10:00 - 13:30:00

The photo-catalytic degradation of Rhodamine-B(RhB) dye was successfully achieved using different heteropolyoxometalate phosphonium ionic liquids(PWPILs) under visble light irradiation. A series of heteropolyoxometalate phosphonium ionic liquids were first prepared from phosphotungstic acid and different quaternary phosphonium salts trihexyl(tetradecyl)phosphonium bromide, tributyl(octyl)phosphonium bromide and tetrabutyl phosphonium bromide with different molar ratioa. The melting points of PWPILs, all lower than 100°C,were characterized by differential scanning calorimeter(DSC). The results show that the characterized peaks of phosphotungstic anion in IR spectra of PWPILs are consistent with that of phosphotungstic acid. This indicates that the quaternary phosphonium cations do not change the Keggin structure of heteropoly acid anion. The influences of experimental parameters, such as the carbon chain length in phosphoniums, the number of free hydrogen in phosphotungstic anions ,addition of different catalysts, irradiation time and the concentration of RhB solution on the photo-catalytic degradation of RhB were investigated. The photo-degradation efficiency was characterized by decolourization rate of RhB. The results indicated that $[P(C_{14}H_{29}(C_6H_{13})_3]_3[PW_{12}O_{40}]$ showed the best photo-catalytic degradation efficiency.
The RhB solution (2×10-4mol/L) was almost decomposed catalyzed by $[P(C_{14}H_{29})(C_6H_{13})_3]_3[PW_{12}O_{40}]$ un der UV-Vis light irradiation for 3 hours. The degradation rate was more than 95%.

(478d) Metal-Containing Ionic Liquid As An Effective Catalyst for the Degradation of Poly(ethylene terephthalate) in Ethylene Glycol

Lu, X., Institute of Process Engineering, Chinese Academy of Sciences
Wang, Q., Institute of Process Engineering, Chinese Academy of Sciences
Zhang, S., Institute of Process Engineering, Chinese Academy of Sciences
Originally presented on: 11/6/2013 13:30:00 - 13:50:00

The vast emissions of poly(ethylene terephthalate) (PET) wastes have caused serious environmental pollution and resource waste. Its effectively recycling has become a global problem which must be urgently solved.

Chemical recycling is an attractive process because PET can be degraded into monomers which can be used as raw materials to produce virgin plastic products. Glycolysis, as an important chemical recycling method, has received extensive concern due to its advantages of less volatile solvents, high conversion of PET, giving monomers and easily continuous production. But there are some drawbacks, such as high temperature, high pressure, low reaction rate and difficult separation between catalyst and product. Many researchers devoted to developing new catalysts to overcome the drawbacks above. Ionic liquids have been reported that could effectively catalyze the degradation of PET to give BHET monomers with high purity and make the reaction condition mild. But the conversion of PET and the selectivity of monomer bis(2-hydroxyethyl terephthalate)(BHET) are both low, and the reaction rate is slow. Therefore, the development of new catalysts is still critical.

It has been found that synergetic effect of catalysts can speed up a chemical reaction or make reaction condition mild.[1,2] Yamamoto et al. found that combinations of organic and inorganic molecules bearing acidic groups could result in higher catalytic activity than individual catalysts for acid-catalyzed reactions.[2] Wang et al. and Zhou et al. found that the synergetic effect of cation and anion of ionic liquid catalyst could make the degradation condition of PET milder.[3,4] Therefore, we infer that a new catalyst containing two active components maybe has highly catalytic activity.

In this study, a metal-containing ionic liquid catalyst including two active components was synthesized and used for the PET degradation in ethylene glycol. The results showed that the synthesized catalyst has better low-temperature catalytic activity. The conversion of PET and the selectivity of BHET are 100 % and 78.7 % respectively at 175 Â°C for 1.25 h. The high catalytic activity is mainly attributed to the synergetic effect of two components. Moreover, this catalyst has the advantages of mild reaction condition, fast reaction rate, products being easily separated from catalyst and being of high purity.

[1] P. R. Schreiner. Science, 2010, 327, 965.

[2] H. Yamamoto and K. Futatsugi. Angew. Chem. Int. Ed., 2005, 44, 1924.

[3] H. Wang, R. Y. Yan, Z. X. Li, X. P. Zhang and S. J. Zhang. Catal. Commun., 2010, 11, 763.

[4] X. Y. Zhou, X. M. Lu, Q. Wang, M. L. Zhu and Z. X. Li. Pure Appl. Chem., 2012, 84, 789.

(478e) Anaerobic Digestion Technology:The Energy,Environment and Health of Developing Nations

Adetule, O., Cornell University

Originally presented on: 11/6/2013 13:50:00 - 14:10:00

Anaerobic digestion is one of the most cost effective and established technologies available for managing organic wastes for energy production,environmental management and pathogen reduction. In developing countries such as Nigeria, the application of this technology at household level as a method for gas generation and safe organic waste management has not been adequately studied. In light of the above, a pilot scale digester was designed for the production of Biogas which at the same time manages waste and reduced pathogens through anaerobic digestion process. The pilot scale digester was 10 litres and manages 9 litres of waste.The digester was fed with different types of waste:Cattle manure, Chicken litter, Rumen of cattle and Water hyacinth. After some weeks, the wastes produced Biogas, reduced pathogens and reduced the amount of wastes in the environment. A prototype digester was later designed and constructed which can accommodate more waste and produced more biogas. The prototype digester is a $5m^3$ underground fixed dome biodigester made of concrete and was designed, constructed and used for digesting cattle dung and water hyacinth. The digester was operated by mixing two parts of cattle dung (900kg) and one part of chopped water hyacinth (100kg) in 2,000 liters of water for two weeks following standard procedures. The gas produced was collected and piped to a kitchen for household use. It was also observed that there was reduction in pathogens and sanitization of the environment through waste management. This study shows that anaerobic digestion can be maximized for waste reduction, energy production,pathogen reduction and conversion of green house gases into usable forms.

(478f) Phosphorus Recovery From Municipal Wastewater Treatment Sludge

Lamichhane Upadhyaya, K., Mississippi State University
French, W. T., Mississippi State University
Hernandez, R., University of Louisiana at Lafayette

Originally presented on: 11/6/2013 14:10:00 - 14:30:00

Phosphorus recovery from municipal wastewater treatment sludge

Kamal Lamichhane Upadhyaya[1], W. Todd French[1] and Rafael Hernandez[2]

[1]Dave C. Swalm School of Chemical Engineering, Mississippi State University, MS, 39762, USA

[2]Department of Chemical Engineering, University of Louisiana at Lafayette, Lafayette, LA, 70503, USA

Abstract

Phosphorus is an essential nutrient for plants, animals and humans. Phosphorus is primarily used in agriculture, as it is one of the important components in fertilizer. Besides, phosphorus is also used in laundry detergent, matchstick, steel production and pesticides. The United States phosphate reserves are declining, which could lead to phosphorus shortage in near future. The United States now imports 10% of its phosphorus supply after exporting for several decades. Hence, there is need to search for renewable and sustainable source of phosphorus such as municipal wastewater sludge. Municipal wastewater sludge contains up to 10 wt% phosphorus and recovering these could significantly decrease the phosphorus shortage in the United States. In this study, the potential of phosphorus recovery from primary and secondary sludge through acidic and basic leaching processes will be investigated. The wastewater treatment sludge will be incinerated to obtain sewage sludge ash (SSA) and subjected to acidic and basic leaching at different acid and base concentration and pH levels. After leaching process, the supernatant will be separated from residue through filtration. Phosphorus will then be recovered from the supernatant mixture through the addition of calcium chloride in basic medium. The recovered phosphorus ($Ca_5(PO_4)_3OH$) has various agricultural and industrial usage as it resembles the phosphate mineral ore (apatite).

(478g) Techno- Economic Analysis of Wastewater Biosolids Gasification

Lumley, N. P. G., Colorado School of Mines
Porter, J. M., Colorado School of Mines
Originally presented on: 11/6/2013 14:30:00 - 14:50:00

Wastewater treatment in the U.S. produces over eight million dry tons of sludge per year. Wastewater sludge in its raw form is a noxious dilute suspension of organic matter in up to 95% water. Management and disposal is an expensive challenge for wastewater treatment plants (WWTPs), where the conventional approach to sludge management is to dewater the material to varying degrees and dispose by landfilling or application as a soil amendment. Such an approach requires energy and financial expense both in drying and ultimate disposal. Dried wastewater sludge has a lower heating value (LHV) close to that of low- rank coal. Thermochemical conversion of wastewater sludge to energy and/ or fuel product may allow for a more value- added approach to sludge management in addition to concentrating and stabilizing the waste stream.

Many WWTPs currently valorize a portion of the energy in sludge via anaerobic digestion. The biological process allows bacteria to operate on dilute sludge in an anaerobic environment producing a combustible bio- gas. Anaerobic digestion is an established technology but allows for only about 20% recovery of the energy in sludge. Anaerobic digestion does not significantly reduce the amount of sludge requiring downstream disposal. Additionally, bio -gas contaminants, especially siloxanes, are a problem for downstream combustion apparatus.

Thermochemical conversion (TCC) technologies subject sludge to chemical processes at elevated temperatures producing heat and/or combustible fuel products. TCCs applicable to wastewater treatment include direct combustion, pyrolysis, and gasification. TCC offers a great advantage over digestion by allowing nearly full utilization of sludge energy and a similarly great reduction in disposable waste volume. Each of these TCCs has been successfully applied to biomass. Wastewater sludge however is fundamentally different from most biomass sources. Raw wastewater sludge may be up to 95% water and 30% ash (dry basis). Balancing moisture removal with reactor operating conditions is the most important consideration in whether TCCs can be net energy- and net economically-positive. Ash concentration and composition present certain challenges to reactor design. Additionally, wastewater sludge organic composition can be variable leading to uncertainty in fuel output.

This work reviews the current state of practice of various TCCs with application to sewage sludge. We discuss technical and economic challenges restricting greater implementation of TCC within the wastewater process. A thermodynamic model of a sludge gasification- electric generator system is built using performance characteristics of a generic, commercially available small biomass gasification platform. We perform energy and techno- economic analyses on sludge gasification to determine lower- limit WWTP capacity and as- delivered sludge moisture content for both air- blown and steam- blown gasification processes.

(513a) Educational Needs in Particle Technology

Jacob, K., The Dow Chemical Company

Originally presented on: 11/6/2013 12:30:00 - 12:45:00

Bulk solids are ubiquitous in our lives today - take a look at the room around you, nearly everything has a significant particle technology component. Consequently, it is no surprise that a new undergrad entering industry has approximately an 80% chance of encountering or dealing with particles in their first project. Yet he or she is significantly ill-equipped to handle the demands of solids processing/particle technology. Merrow, both in the mid-80s and in 2001, has highlighted the significant challenges engineers face when dealing with solids due to a lack of education in the area and the resulting consequences: considerable downtime, lost production, start-up difficulties, etc. Industry has responded with post graduate training for their engineers or with the creation of industrial engineering science laboratories dedicated to particle technology that are on equal footing with more traditional chemical engineering disciplines such as reaction engineering, separations, fluid mechanics, heat transfer,etc. It is very important, if not imperitive, that solids processing/particle technology be incorporated into the chemical engineering curriculum. We will examine the challenges the chemical engineering community faces to meet this need.

(513b) Teaching Mixing to Undergraduate Chemical Engineers

Grenville, R. K., Philadelphia Mixing Solutions Ltd.

Originally presented on: 11/6/2013 12:45:00 - 13:00:00

Chemical engineers take raw materials and mix them so that they react to form valuable products and by-products.Â The products and by-products are then separated.Â While these steps are being performed the components may be heated or cooled as necessary.Â Of these four basic operations, three are covered in the undergraduate chemical engineering curriculum.

Smith (Trans. IChemE, 1990) estimated that the US chemical industry loses approximately $ 1 billion a year due to problems associated with poor understanding of the role of mixing on the processes being operated.Â Given the importance of mixing as a discipline, it is surprising that there are very few universities that offer a formal course on the subject.

After graduation engineers can take continuing education courses or webinars on mixing organized by, among others, the Center for Professional Advancement, the American Institute of Chemical Engineers and the British Hydromechanics Research Group.Â Some larger companies that employ mixing specialists offer in-house seminars on mixing to their employees.

If a formal course on mixing cannot be included in the curriculum due to, for example time constraints, a single lecture could be included in other courses, such as fluid mechanics, reaction engineering and transfer processes, that would show how the concepts being taught can be extended to take the role of mixing into account.Â These modules could be prepared as webinars aimed at undergraduate students rather than those offering continuing education for engineers in industry.

(513c) Addressing Educational Gaps for Process Safety in Both Academia and Industry

Vaughen, B. K., BakerRisk

Originally presented on: 11/6/2013 13:00:00 - 13:15:00

Members of AIChE's Safety and Health Division (S&H) have been working for decades with AIChE's Center for Chemical Process Safety (CCPS) to develop process safety-related continuing education materials for practicing engineers. To help address the needs of chemical engineering educators, members of the S&H Division help bridge the gap between industry and academia through the Safety and Chemical Engineering Education (SAChE) committee. One of the SAChE committee's goals is to help instructors who don't have extensive process safety experience prepare for and teach chemical process safety examples and lessons in their classes. Extensive process safety guideline books, webinars and "safety certificates" (available through SAChE) have been developed and are currently in use today. Integrating process safety system principles, such as inherently safer design and process hazards analyses, into an already overloaded chemical engineering curricula has been one of the SAChE committee's challenges, and is a "hot topic" today, especially now that the Accreditation Board of Engineering and Technology (ABET) added a "process safety" expectation into its audit protocol in the Fall of 2012.

(513d) Engineering Risk Rountable

Matonis, D., Flucomp Enterprises, Inc.

Originally presented on: 11/6/2013 13:15:00 - 13:30:00

The inherent complexity in manufacturing operations through construction have put more pressure on companies to focus on risk management beyond regulations and standards compliance. However, many individual facilities use a multitude of systems and methods and descriptions to define and assess engineering project and construction risks along with different models to score the severity and likelihood of risk and to control it. This creates isolated pockets of risk knowledge, and impedes the ability to accurately share lessons learned and best practices effectively within one company let alone across an industry. The cost of capital, resources and personnel will continue to rise while regulatory requirements and sustainability initiatives become more stringent, integrated, and expensive. All of these factors create a strong need for strategic operational risk and information management uniformity and education.

(513e) Training Chemical Engineers for a Career in Process Development

Hollar, W., SAUDI BASIC INDUSTRIES CORPORATION

Originally presented on: 11/6/2013 13:30:00 - 13:45:00

Training undergraduate chemical engineers for careers in process development represents a unique challenge. Traditional chemical engineering unit operations remain the basis for much of the chemical industry both in the United States and globally. The expansion of chemical engineering into non-traditional areas such as biochemical processes, materials and interdisciplinary areas such as nanotechnology has placed additional driving forces on chemical engineering departments to provide basic training in these areas. This can potentially come at the cost of reducing the depth of training in areas such as process control, process design and unit operations. This comes at a time when the need for young chemical engineers to develop competitive, sustainable processes in energy production and delivery, globally competitive commodity chemicals, and unique high value specialty chemicals continues to rise.

Careers in process development require the conflicting requirements of both depth and breadth in the field of chemical engineering for optimal performance. While much of the training occurs on the job, the education at the university represents the solid base upon which breadth and depth is built. Some opportunities for increasing the level and effectiveness of the interaction between industry and academia are proposed:

Invited lectures by industry chemical engineers in relevant class topics

Heavier participation of industry in classes such as senior design courses

Alternatives for in-depth training in specific chemical engineering topics

Increased funding and opportunities for student research and internships in industry

(513f) Responding to the Needs of Industry: Perspectives From the Education Division
Justice, F.,
Originally presented on: 11/6/2013 13:45:00 - 14:00:00

The membership of the Education Division, by definition have a keen interest is building into the undergraduate curricula those elements that will make one of our products ,"new Chemical Engineering practitioners", valuable to the industries that employ them. Like any business, the 'feedback' from the marketplace is one of the truest inputs which when accurately measured can point the way to continuous improvement activities that assure the graduates are both well-grounded in fundamentals and able to quickly come up to speed in emerging areas.

Getting and verifying this market place feedback and communicating it to Education professionals is crucial to evolutionary improvements of the curricula and teaching methods.

(513g) Responding to the Needs of Industry: Nuclear Engineering Division

Olson, J., Process Engineering Design Manager

Originally presented on: 11/6/2013 14:00:00 - 14:15:00

The nuclear industry needs engineers to maintain and design the future facilities for both power generation and waste cleanup. Chemical engineers, due to their unique background and training, are highly sought-after resources in the nuclear industry. Bechtel categorizes engineers as Process/Chemical Engineers, or Nuclear Engineers. Both work on nuclear industry jobs, but their focus is different.

Process/chemical engineers apply the principles of chemistry to solve problems involving the production or use of chemicals and bio-chemicals. They design equipment and processes for large-scale chemical manufacturing/cleanup, plan and test methods of manufacturing products and treating byproducts, and supervise production. In BSII, Process/Chemical Engineers are responsible for developing Process Flow Diagrams to include process calculations and producing System Design Descriptions. They prepare equipment process designs and are responsible engineers in the purchase of process equipment. Often, they are responsible for engineering studies up to and including pilot scale testing.

Nuclear engineers research and develop the processes, instruments, and systems used to derive benefits from nuclear energy and radiation. They design, develop, monitor, and operate nuclear plants to generate power. They may work on the nuclear fuel cycle—the production, handling, and use of nuclear fuel and the safe disposal of waste produced by the generation of nuclear energy—or on the development of fusion energy. Some specialize in the development of nuclear power sources for naval vessels or spacecraft; others find industrial and medical uses for radioactive materials, as in equipment used to diagnose and treat medical problems. In BSII, Nuclear Engineers develop the calculations associated with nuclear facilities and provide direct support to other engineering disciplines for nuclear related matters.

(513h) Responding to the Needs of Industry: Sustainable Energy Forum

Knox, P., Eco Chem Strategies

Originally presented on: 11/6/2013 14:15:00 - 14:30:00

This is an abstract for the Needs of Industry session, with perspective from the Sustainable Energy Forum. It is an invited talk. This paper will discuss the skills level of graduating B.S. chemical engineers as it relates to the needs of sustainable energy. A panel discussion will follow all of the invited talks.

(513i) Panel Discussion: Responding To Industry's Needs

Rau, N., Sackett Solutions & Innovation
Visco, D. P. Jr., The University of Akron
Originally presented on: 11/6/2013 14:30:00 - 15:00:00

This is a panel discussion featuring all of the authors who have presented in this session, moderated by the session chair and the session co-chair. Audience members will have an opportunity to ask questions of the presenters. Nine different speakers will present and represent the leadership, past or present, from their division or forum.

(521a) Characterization of Bio-Oil, Bio-Char from Biomass Fast Pyrolysis and Bio-Char Application into Soil

Wang, Y.,
Liu, R.,
Yin, R.,
Wu, L.,
Originally presented on: 11/6/2013 12:30:00 - 12:42:00

Fast pyrolysis of biomass was carried out in a fluidized bed reactor. Three kinds of products including bio-oil, biochar and non-condensible gases were obtained. Characterization of bio-oil, bio-char from biomass fast pyrolysis and bio-char application into soil were investigated. Bio-oil is a clean, cost effective liquid alternative to replace fossil fuel for 'green power' generation, transportation and district heating. However, the properties and chemical composition of bio-oil are distinct from petroleum fuel. In general, due to its relatively high water and oxygen content, bio-oil is acidic and not stable when heating. Components of bio-oil from fast pyrolysis of rice hull and their relative mass contents were detected by GC–MS method. It is shown that fast pyrolysis is a complex process. 112 types of chemical compounds were detected. The ash and fixed carbon content increased and volatile component content decreased for all the three feedstock including elm wood, pine wood and rice hull after pyrolysis. The rice hull biochar had the highest ash content and the lowest fixed carbon and volatile component, which was 50.94%, 33.35% and 13.11%, respectively. The pH value of biochar was from neutral to alkalinity. Of the three kinds of biochar, the pH of pin wood biochar(9.06) was higher than that of rice hull biochar(7.49) and elm wood biochar(6.35). Therefore, rice hull biochar and pine wood biochar could be applied to acid soil improvement. Rice hull biochar had the highest specific surface area (BET) (78.15 m3/g) and apparent densities (418.3 kg/m3), and SEM analysis also confirmed that rice hull biochar had the most micropores. Fourier transform-infrared (FT-IR) demonstrated that the functional groups of three kinds of feedstock were similar, and also similar of three biochar except that rice hull biochar had extra arene functional groups. Results of two-month soil experiment showed that rice hull biochar(4%) could increase soil nutrition including N, K, Mg and Mn, especially for K content of soil which increased from 0.055 to 0.417mg/kg. The soil carbon increased 1.69 times, and the pH value increased from 3.33 to 3.62, but there was no significant effect to phosphorus. The most important thing is that rice hull biochar reduced the Pb and Al availability to some extent, so it could be used to improve the soil quality that was contaminated by heavy metals. Although application biochar into soil has many advantages, a long-term field experiment and the effect of production of biochar on environment need to be estimated before application of biochar in a large scale.

(521b) Process Development and Optimization of Cellulosic Derived Sugars from Biomass for Biofuels and Chemicals Applications

Parekh, S., Sweet Water Energy

Originally presented on: 11/6/2013 12:42:00 - 12:54:00

The recent volatility of petroleum prices is stimulating the biorefining industry to diversify its raw material feedstock, using a broad-based sugar platform technology. Corn silage and energy crops such as sweet sorghum are widely recognized as the promising high-volume, low-cost lignocellulosic feedstocks from which to derive C5- and C6-rich sugar streams that can be bioconverted into biofuels and chemicals.

This presentation will review Sweetwater Energy's overall approach and concept of developing a process based on a decentralize model, whereby a modular, feedstock-flexible bioprocessing machine pretreats and processes ligocellulosic biomass, and generates adequate quantities of C5 and C6 sugar-rich streams at the biomass-generation site. The sugars are then transferred to a regional refining site for the production of biofuels. This talk will highlight: 1) the importance of biomass harvesting and storage technologies from the field to the bioprocessing site, 2) characterization of the physical and chemical properties that enable optimum bioconversion process, 3) specific pretreatment conditions and various process strategies for deriving streams rich in C6 and C5 sugars for biofuels production, and 4) review recent advances made at Sweetwater Energy in sugar platform technologies, 5) process scale up thereby accelerating process commercialization pretreating a wide range fof biomass into cellulosic sugars that can be used for various applications in ethanol chemicals and advanced fuels, such as bio-jet fuel.

(521c) Enzymatic Hydrolysis of Old Corrugated Cardboard (OCC) Fines from Recycled Linerboard Mill Waste Rejects

Min, B. C., SUNY College of Environmental Science and Forestry
Bhayani, B., Empire State Paper Research Institute
Ramarao, B. V., ESPRI, SUNY College of Environmental Science and Forestry
Originally presented on: 11/6/2013 12:54:00 - 13:06:00

During the manufacture of linerboard from recycled OCC (old corrugated cartons), a significant fraction of the fibers are broken into small fine particles. These fines are rejected into the solid waste stream placing a cost for landfilling and posing environmental problems. The fines rejects are composed primarily of cellulose which can be hydrolyzed into sugars for possible fermentation into biofuels or other products. These fines have some advantages such as low cost and without any necessity of complex pretreatment processes characteristic to reduce lignocellulosic biomass. The enzymatic hydrolysis of reject fines from a recycled OCC mill was studied for producing sugars and the methods of reducing costs were considered. Three commercially available enzyme mixtures were tested. Cellulase from Trichoderma Reesei was found to be the most effective at loading levels of 5 – 100 FPU (/g of oven dry mass). The maximum hydrolysis yield was 45% sugar (g/g-OD fines) with 50 FPU. Since the fines rejects contain fillers (upto 30% by mass), the enzyme loading was lowered by using surfactants to cover the mineral surfaces. Reductions of the enzyme dosage were observed to obtain high hydrolysis yield using non-ionic surfactants (Tween 80, CTAB) or different pH buffer. As a result, Tween 80 reduced enzyme requirement to 30 FPU and Combination Tween 80 and pH 4 sodium acetate buffer increased efficiency of low enzyme dosage.

(521d) Effects of Fuel Properties on Natural Downward Smoldering of Piled Biomass Powder: Experimental Investigation

He, F.,
Gao, Z.,
Luo, B.,
Li, Z., Shandong University of Technology
Yi, W., Shandong University of Technology
Originally presented on: 11/6/2013 13:06:00 - 13:18:00

Effects of fuel properties on natural downward smoldering of piled biomass powder: experimental investigation

Fang He, Zhenqiang Gao, Bin Luo, Zhihe Li, Weiming Yi

Shandong University of Technology, Zibo, Shandong, 255049 P.R. China

Abstract: In order to obtain data for modeling validation of one-dimensional smoldering and combustion of biomass, effects of fuel type, moisture content, and particle size on natural downward smoldering of biomass powder have been investigated experimentally. A cylindrical reactor (inner size Φ26 cm Ã— 22 cm) was built and powders of corn stalk, pine trunk, pyrolysis char and activated char from corn stalk were prepared.

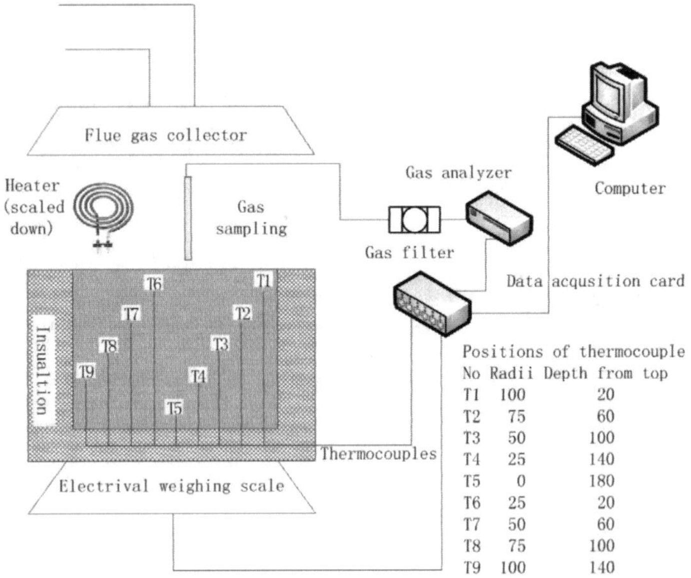

Fig. 1 Schematic of the experimental set-up for the natural downward smoldering

Smoldering characteristics of the four materials, different moisture contents (between 0% and 30%, wet basis, corn stalk powder) and different particle sizes (between 20 and 100 mesh, dry corn stalk powder) were examined. Results showed that: 1) The maximum temperature in the fuel bed is only slightly affected by fuel type and particle size. It increases gradually for original biomass and decreases slowly for chars with the development of the process. 2□Propagation velocity of the char oxidation front is dominated by carbon density and ash content. And it is almost not affected by moisture content and particle size. 3) There is always a drying front in the fuel bed during smoldering of hydrocarbons even if the original material is dry. Propagation velocity of the drying front is affected by moisture content significantly, decreases from more than 10 to 3 times of char oxidation velocity with the increase of moisture content from 0 to 20%. 4) Cracks inside the fuel bed of are dominated by fuel type and particle size.

Typical results:

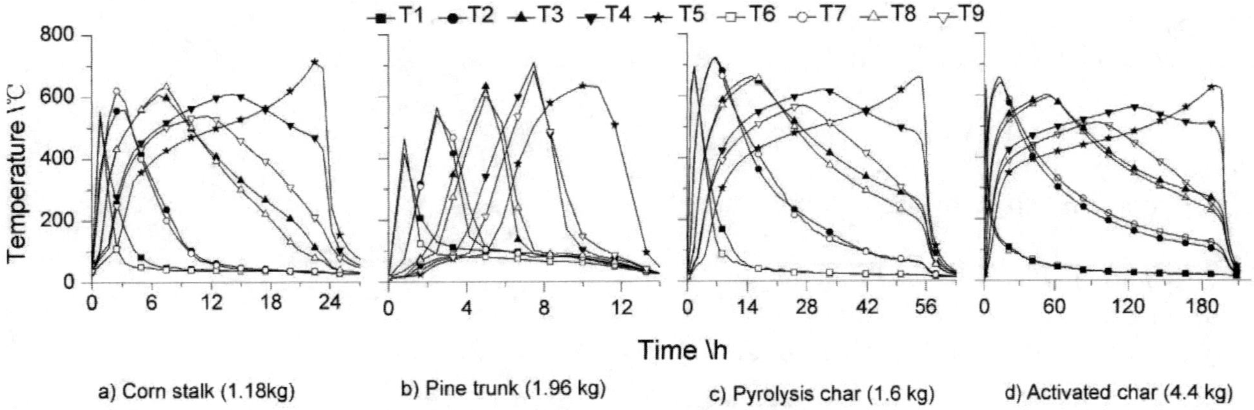

Fig. 2 Effects of the fuel type on the temperature of smoldering

Temperature histories of smoldering of the four fuels are presented in Fig. 2. Maximum temperatures from thermocouples are between 450 and 750 ℃, varies with their depths.

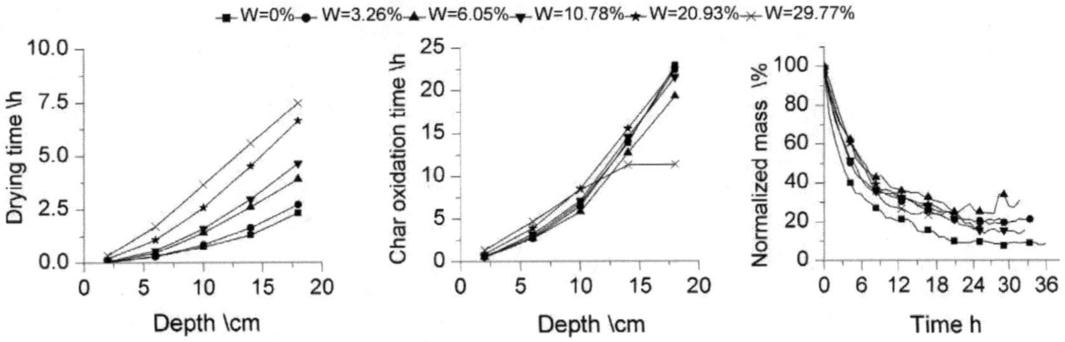

Fig. 3 Effects of moisture content on propagation and mass loss rate of smoldering

The time it takes for the thermocouples to reach the given temperatures□101 ℃for drying and the maximum for char oxidation) at moisture content between 0 and 30 % is presented in Fig. 3. It shows that the drying time increases significantly with the increasing moisture content. However, there is no obvious regularity between the char oxidation time and the moisture content.

Highlight:

1. Natural downward smoldering of four materials, different moisture contents, and different particle sizes were investigated.

2. Propagation velocity of the char oxidation front differs significantly from that of the drying front.

3. Carbon density and ash content of fuel dominate propagation velocity of the char oxidation front.

4. Fuel bed of finer material is apt to have bigger cracks.

Keywords: fuel type, moisture content, particle size, biomass, smoldering

(521e) Experimental Study of Bio-oil Production from Corn Stalk Using a Fluidized Bed under Hot Flue Atmosphere

Li, Z., Shandong University of Technology
Niu, W.,
Yi, W., Shandong University of Technology
Li, Y.,
Fu, P., Shandong University of Technology
Zhang, D.,
Wang, J., Shandong University of Technology
Bai, X., Shandong University of Technology
Originally presented on: 11/6/2013 13:18:00 - 13:30:00

Experimental study of Bio-oil Production from Corn Stalk using a Fluidized bed under hot flue atmosphere

Zhihe Li[a], Weisheng Niu[b], Weiming Yi[a]*, Yongjun Li[a], Peng Fu[a], Deli Zhang [a], Jing Wang[a], Xueyuan Bai[a] *

[a]*School of Agricultural and Food Engineering, Shandong University of Technology, Zibo 255049, China*

[b]*School of Engineering, Shenyang Agricultural University,Shenyang110866, China*

*Corresponding author. Tel.: +86 533 2786308; Fax: +86 533 2786158

Abstract

In order to investigate the effects of pyrolysis temperature on the product yields, Â fast pyrolysis experiments for bio-oil production have been conducted on corn stalk using a novel fluidized bed reactor under hot flue atmosphere at the selected temperature range of 450-520℃. During experiments, un-condensable gases were detected on-line with Gasboard-3100. The bio-oil obtained under the experimental conditions that gave maximum oil yield were used for further characterization.The chemical compositions of the bio-oil were determined by GC-MS. Physical analysis of the bio-oil such as pH value, viscosity and calorific value were determined with PHS-2F pH meter (Shanghai Precision and Science Instrument Co., LTD), NDJ-1 rotational viscometer (Shanghai Changyi Gealogical Instruments Co., Ltd) and WZR-1T oxygen combustion bomb (Changsha Bente Apparatus), respectively.

The results show that maximum bio-oil yield of 54.1wt% is achievedd at 500℃. The main components of un-condensable gaseous mixture includes Methane, Hydrogen, Carbon dioxide and Carbon monoxide. Bio-oil is a complex compound which contains many chemical species and it is an acidic, high viscosity, low heat value and low volatility liquid in comparison with diesel fuel. The properties change rapidly during storage condition. Its utilization as fuels requires some form of upgrading to improve storage stability and heat value and reduce the water content, viscosity and acidity.

Key words: Pyrolysis; Fluidized bed, Corn stalk; Bio-oil; Flue

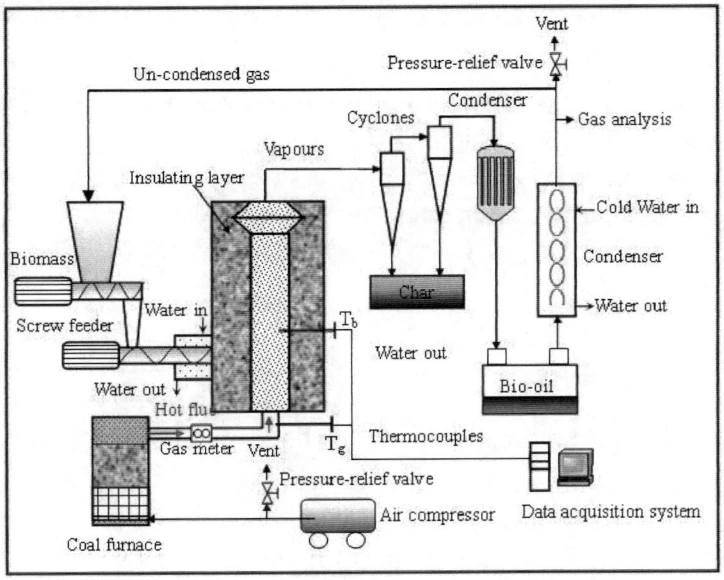

Fig. 1. Schematic diagram of the fluidized bed experimental system

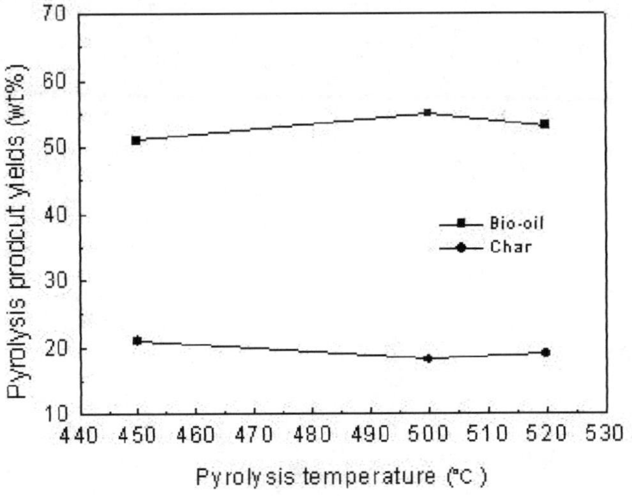

Fig. 2 Product yields of corn stalk at selected pyrolysis temperature

(521f) Quantum Chemistry Study on Adsorption of Gases from Biomass Pyrolysis on CaO Surface

Zhao, B., Energy Research Institute of Shandong Academy of Sciences
Zhang, X., Energy Research Institute of Shandong Academy of Sciences
Chen, L., Energy Research Institute of Shandong Academy of Sciences
Yi, X.,
Tian, Y.,
Chen, G.,
Originally presented on: 11/6/2013 13:30:00 - 13:42:00

Biomass pyrolysis/gasification with in-situ CO_2 absorption is a promising technology to produce the rich-hydrogen fuel gas, which can be used for wide range of applications as well as the operation of fuel cell, the generation of synthetic natural gas and hydrogen, or the heat supply through combustion. It is necessary to investigate the effects of the reaction atmosphere on CaO carbonation performance due to its role to enhance H_2 production reactions. The adsorption of CO, O_2, H_2O(g) , CO_2 , CH_4 and H_2 on CaO surface were studied by quantum chemistry method in this paper. The corresponding microscopic parameters, such as adsorption energies, adsorption distances, adsorption bond orders and changes of net charge of gases on CaO surface were calculated with the soft of material studio. The curves of binding energies between gases and CaO surface were fitted well using the mathematic function. It was indicated that the adsorption of CO_2 on CaO surface was the stronger than those of H_2O, CO, O_2, CH_4 and H_2.

(534a) Plate Food Waste: A Valuable Source of Energy

Jung, S., Iowa State University

Originally presented on: 11/6/2013 15:15:00 - 15:35:00

Post-consumer food waste, because of its quantity and composition, could be a good feedstock for biofuel production. The effect of glucan loading (2 and 6%), addition of an enzyme cocktail, pretreatment (grinding/drying or sterilizing), and food waste composition (breakfast vs. lunch, presence of used napkins) on ethanol concentration was reported. The potential for sorting and converting used napkins into ethanol was also assessed. At a 2% glucan content, ethanol concentration of up to 16 g/l was obtained after saccharification/fermentation of breakfast and lunch food waste. Increasing glucan content to 6% significantly increased ethanol concentration to 54 g/L, while there was no significant effect of food waste composition on carbohydrate conversion efficiency. When converting ethanol concentration from g/L to L/g of waste, concentrations of 0.06, 0.06 and 0.40 were obtained for food waste, food waste w/o napkins, and used napkins, respectively, identifying used napkins as an excellent source of bioethanol production.

(534b) Biofuel Production and Kinetics Study of Catalytic Microwave Pyrolysis of Douglas Fir Pellet Over Activated Carbon Supported Metal Catalyst

Bu, Q., Washington State University
Lei, H., Washington State University
Wang, L., Washington State University
Tang, J.,
Originally presented on: 11/6/2013 15:35:00 - 15:55:00

The presented study is aiming to improve the quality of bio-oils by catalytic upgrading of pyrolysis oil from microwave pyrolysis of biomass using activated carbon (AC) supported metal catalyst. A central composite experimental design (CCD) was used to optimize the reaction condition of this process. The effects of reaction temperature and reaction time on product yields were investigated. GC/MS analysis indicated that the main chemical compounds of bio-oils were furans, phenols, guaiacols, and ketones/ethers which were in the range of 80-87% of bio-oils for most of the samples. The ketones/ethers mainly composed by 2-cyclopen-1-one and straight chain ethers were increased up to about 38% of bio-oil, and the phenols increased significantly accompanying with the decrease of guaiacols in comparison with raw biomass pyrolysis. Thermogrametric analysis (TGA) was used to analyze the thermal behavior or Douglas fir pyrolysis, and the kinetic parameters were investigated using thermogrametric analysis (TGA) data and compared with catalytic microwave pyrolysis of Douglas fir in this process, the activation energy of Douglas fir pyrolysis in this process was remarkable lower than conventional TGA pyrolysis of Douglas fir. And the reaction mechanism of this process was analyzed.

(534c) Molecular-Level Modeling of Municipal Solid Waste Gasification

Horton, S. R., University of Delaware
Zhang, Y., Air Products and Chemicals, Inc.
Bennett, C. A., University of Delaware
Klein, M. T., University of Delaware
Petrocelli, F., Air Products and Chemicals, Inc
Originally presented on: 11/6/2013 15:55:00 - 16:15:00

Molecular-level Modeling of Municipal Solid Waste Gasification

Scott R. Horton*, Craig A. Bennett, Michael T. Klein

Department of Chemical & Biomolecular Engineering, University of Delaware

Yu Zhang, Francis P. Petrocelli

Air Products and Chemicals, Inc.

In 2010, there was 250 million tons of municipal solid waste (MSW) in the United States. Over half of this waste ended up in landfills which fail to recover the energy stored within MSW. The most basic waste-to-energy technology is incineration. While this process recovers energy, it also poses problems to the environment due to SO_x, NO_x, dioxin, and furan emissions. Plasma-arc gasification has nonstoichiometric amounts of oxygen and reduces the production of these harmful oxygenates while recovering energy from waste via syngas production. Furthermore, due to the extremely high temperatures of the plasma torch, the final by-product is a salable glass-like slag. The objective of this study is to aid process design by developing a molecular-level kinetics model of plasma-arc gasification.

The MSW is typically described at the lumped level, i.e., biomass, food waste, and plastics. In the present work, each of these lumped components was described at the molecular-level. Biomass was modeled as cellulose, hemicellulose, and lignin. In addition to these molecules, lipids, common amino acids, and sugars were used to model food waste. The modeled plastics were polyethylene, polyethylene terephthalate, and polyvinyl chloride. These feeds were all represented at the molecular level using Flory statistics for nearly linear polymers and lattice theory for lignin.

The feed stream was converted into a reaction network using an in-house software tool, the Interactive Network Generator (INGen). There were two major categories of reactions: devolatilization/pyrolysis and gasification. The major pyrolysis reaction families were thermal cracking, decarboxylation, and decarbonylation. For gasification, the dominant reaction type was reaction with oxygen. The generated reaction network was solved and optimized using another in-house software, the Kinetic Modeling Editor (KME).

For integration with process simulation, the optimized KME model was integrated into ASPEN PLUS using user-defined models. In the case of gasification, the KME model contains reactors in series and rate laws employing linear-free energy relationships for reaction families. The process of KME-Aspen integration was generalized allowing for the automatic implementation of any reaction model into an ASPEN worksheet.

The KME model could also be coupled to a computational fluid dynamics (CFD) solver and take equipment level information into account in the design. Once the products and kinetics of the pyrolysis and gasification reactions are obtained from KME, the reaction network could be converted into one or several global pyrolysis/gasification reactions by kinetic model reduction. The simplified mechanism could be easily integrated into any CFD solver, such as Ansys FLUENT. On the other hand, equipment information obtained from the CFD simulation, such as flow and temperature pattern, provides guidance for choosing reactors in ASPEN PLUS process simulation.

(534d) Performance Analysis of Rdf Gasification in a Two Stage Fluid Bed - Plasma Process

Materazzi, M., University College London

Lettieri, P., University College London

Chapman, C., Advanced Plasma Power

Taylor, R.,

Originally presented on: 11/6/2013 16:15:00 - 16:35:00

PERFORMANCE ANALYSIS OF RDF GASIFICATION IN A TWO STAGE FLUID BED - PLASMA PROCESS

M. MATERAZZI* **, P. LETTIERI*, R. TAYLOR **, C. CHAPMAN **

** Department of Chemical Engineering, University College London, Torrington Place, London WC1E 7JE, UK*

*** Advanced Plasma Power, South Marston Business Park, Swindon, SN3 4DE, UK*

Tar generation and ash disposal represent the strongest barrier for use of single stage fluid bed gasification for waste treatment, whereas sufficing for both is only possible with expensive cleaning systems and further processing. Waste fuels have, in general, physical and chemical characteristics much different from fossil fuels, so that the behaviour of these two feedstock types in high temperature fluid beds is very different one from the other. The high content of ashes and inorganic materials in RDF (Refuse Derived Fuel) dictates the thermal conditions in the fluid bed. A relatively low temperature has to be used to prevent agglomeration and sintering of bed material. For this reason, the gas that is produced by a standard fluid bed gasifier (FBG) has tars and other condensable organic species that are technically difficult and costly to remove. Furthermore, the bottom ash that is generated in the reactor may contain high levels of carbon, heavy metals and organic pollutants which lower the conversion efficiency of the process and limit any secondary usage.

The use of plasma systems has increasingly been applied with thermal waste treatment for its ability to completely decompose the input waste material into a tar-free synthetic gas and an inert, environmentally stable, vitreous material known as slag. The principal advantages that plasma offers to thermal conversion processes, besides the already mentioned tar/ash related issues absence, are a smaller installation size for a given waste throughput, and the use of electricity as energy source, characteristics which permit the technology to treat a wide range of low calorific value materials including liquids and solids. In applying the plasma technology to the gaseous products from a FBG, an advanced two-stage thermal process is able to achieve efficient conversion of the chars and the complex organics to the primary syngas constituents whilst limiting the electrical energy demand of the process (Figure 1).

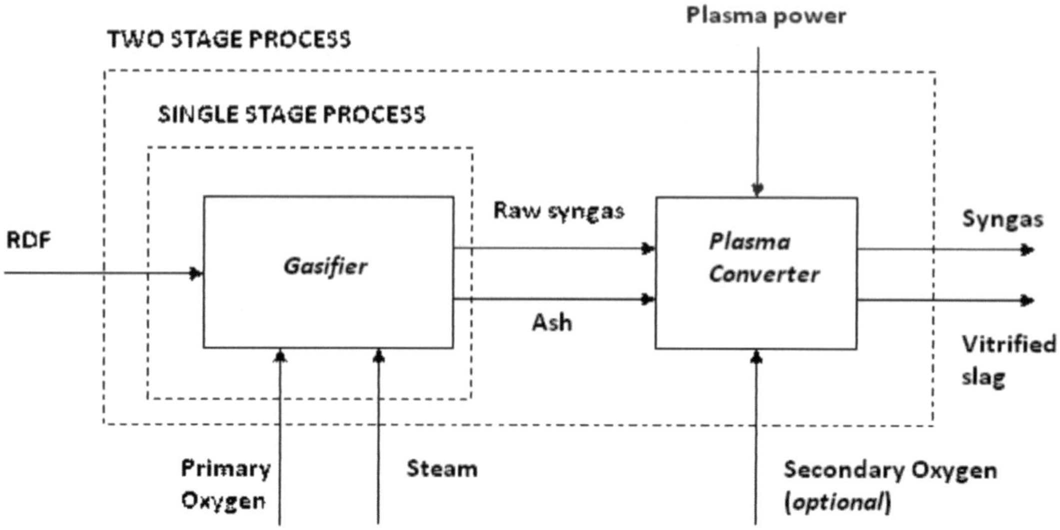

Figure 1: Schematic of Gasplasma two-stage thermal process.

This study focused on the thermodynamic assets of using a two-stage (FBG + plasma) thermal process over the conventional single-stage approach. These include, for example, the fact that the primary thermal waste decomposition is performed in conditions of optimal stoichiometric ratio for the gasification reactants. Furthermore, staging the oxidant injection in two separate intakes significantly improves the efficiency of the system, reducing the plasma power consumption. The aim of the present work is also to determine the effects of the thermal parameters so as to be able to select optimum gasification conditions in a two stage process and to clarify the mechanism of thermal decomposition of carbonaceous materials in the FBG and in the plasma converter.

A flexible model capable of providing reliable quantitative predictions of product yield and composition after the two-stage process has been developed. The method has a systematic structure that embraces atom conservation principles and equilibrium calculation routines, considering all the conversion stages that lead from the initial waste feed to final products. The model was also validated with experimental data from a demonstration plant in Swindon (UK) (Table 1).

	Case 1	Case 2	Case 3	Case 4	Case 5
Description:					
O2/fuel ratio (w/w)	0.51	0.59	0.59	0.79	0.50

Bed temperature (°C)	770	720	795	720	800
Proximate analysis, % (w/w)					
Fixed carbon	6.4	12.2	11.6	8.5	22.8
Volatile matter	59.6	50.2	64.8	47.6	68.0
Ash	19.1	23.2	12.1	8.9	0.5
Moisture	14.9	14.4	11.5	35.0	8.7
Ultimate analysis, % (w/w)					
C	41.0	47.0	43.0	31.5	45.2
H	5.7	6.3	5.6	4.1	6.46
O	17.5	6.9	26.6	19.7	45.38
N	1.2	1.74	0.61	0.4	0.26
S	0.2	0.15	0.25	0.17	0.01
Cl	0.4	0.31	0.34	0.23	0.25
GCV, MJ/kg (dry basis)	22.1	26.4	21.0	21.0	22.0

Table 1: Experimental parameters and characteristics of solid wastes (as received).

The study effectively demonstrated that the two-stage gasification system significantly improves the gas yield of the system and the carbon conversion efficiency, which are crucial in other single stage FBG systems, whilst maintaining high energy performances.

The most important outcome of this analysis is that the important parameters that define the performance of the fluid bed-plasma process are the oxidants (oxygen and/or steam) streams and the plasma power. These parameters are function of the type of waste and the detailed aim of the project. Furthermore, the two-stage performance can be quite accurately predicted by assumption of thermodynamic equilibrium.

Figure 2: Comparison of model results with experiment data from the (left) FBG gasifier; (right) FBG + Plasma converter

There is close correlation between the observed and predicted values of the syngas exiting the plasma converter, although the actual level of CO is still ~1% below the theoretical value while the CO_2 level is 1.5% higher. The noted differences are likely due to flow and RDF composition uncertainties rather than from the predictive model. Given the size and configuration of the demonstration plant and the inherent variability of real RDF uncertainties are to be expected. The layout of syngas heating value (LHV) and cold gas efficiency (CGE) have been calculated and analyzed. It was found out that high LHV and CGE values are maintained in different power and oxygen conditions. The reason is that addition of plasma power into the converter decreases the amount of secondary oxygen required for complete gasification and produces larger amounts of CO and H_2 in the product gas. The optimizing direction for the two-stage process can only be determined after considering the detailed situation on different projects.

(534e) Process Modeling of Continuous Catalytic Gasification As a Waste to Energy Alternative

Tillie, C. F., Cleveland State University
Riedthaler, M. J., Cleveland State University
Reeves, S. A., Cleveland State University
Adhikari, S., Cleveland State University
Gatica, J. E., Cleveland State University
Originally presented on: 11/6/2013 16:35:00 - 16:55:00

Presently, space exploration waste is either carried onboard for the duration of the mission or discarded into a trash module which is sent back into Earth's atmosphere for destruction. On the other hand, with an estimated cost of $10,000 per pound, the cost of delivering mass to locations beyond low earth orbit (LEO) locations is prohibitive.

Since approximately 15 pounds of waste are produced daily by a crew of four during a mission, the intrinsic value of this waste can be greatly enhanced by waste to energy technologies. Moreover, conceptualizing waste gasification as an in-situ alternative for processing of trash is also an efficient route to control waste, while maintaining a healthy habitable environment, during long-duration missions (Santiago-Maldonado et al., 2010).

The gasification of long-chain polymers is hypothesized to follow a mechanism of four combined reactions —two parallel oxidation reactions which are kinetically controlled, accompanied by two series reaction where oxidation products (carbon dioxide and carbon monoxide) are converted into methane and hydrogen. Integrating this process into spaceflight missions would greatly reduce the onboard mass requirements, as these products can be used as propellant fuel. Further, under certain conditions the process could be also be adapted to produce oxygen and water, two essential life support species for space missions.

This paper explores scale-up alternatives to implement low-temperature continuous gasification processes as potential waste management processes. A kinetic model formulated from batch gasification experiments is integrated into a process simulator along with a detailed thermodynamic description.

Emphasis is placed on the preparatory treatment of the waste slurries, the reactor module, and equilibrium stages for recycling reactants and recovering the catalyst. Detailed characterization of low and mid fidelity simulants is presented.

(534f) Performance Evaluation of Accelerated Carbonation of Basic Oxygen Furnace Slag Via a Rotating Packed Bed: Modeling, Analysis and Maximization

Pan, S. Y., National Taiwan University
Chiang, P. C., National Taiwan University
Chen, Y. H., National Taipei University of Technology
Chang, E. E., Taipei Medical University
Tan, C. S., National Tsing Hua University
Originally presented on: 11/6/2013 16:55:00 - 17:15:00

Performance of accelerated carbonation by basic oxygen furnace slag (BOFS) in a rotating packed bed (RPB) was evaluated in this investigation. The design of experiments (DOE) was introduced to evaluate the effect of different operating parameters including reaction time, particle size of BOFS, temperature, liquid-to-solid (L/S) ratio, and rotating speed on carbonation conversion of BOFS. A quadratic response surface was established statistically to yield maximum carbonation conversion based on experimental data. In addition, both the fresh and carbonated BOFS were characterized by the thermogravimetry and differential scanning calorimetry (TG/DSC), X-ray diffraction (XRD), and scanning electron microscopy equipped with X-ray energy dispersive spectroscopy (SEM-XEDS) to confirm the results of accelerated carbonation of BOFS in an RPB.

Furthermore, by combining the process chemistry, reaction kinetics, and mass transfer, a prediction model for carbonation of BOFS within an RPB, with an error less than 10%, was well-developed. The enhancement factors for mass transfer coefficient in case of carbonation of BOFS within an RPB were determined accordingly. In comparison with the developed model and the RSM results, a standard operation procedure (SOP) was proposed for maximizing the carbonation conversion of BOFS. It was thus concluded that the accelerated carbonation of BOFS using an RPB is a promising process for CO_2 capture due to its higher carbonation conversion achieving in a short reaction time.

(534g) Production of Alumina From Fly Ash By a Mild and Cleaner Hydrometallurgical Process

Zhong, L., Institute of Process Engineering, Chinese Academy of Sciences
Duan, D., Institute of Process Engineering, Chinese Academy of Sciences
Zhou, E.,
Han, H.,
Originally presented on: 11/6/2013 17:15:00 - 17:35:00

Fly ash is a large volume solid waste generated during the coal combustion by coal-fired power plants, wherein the valuable alumina content could be up to 30-50%. A mild hydrometallurgical process to produce alumina from fly ash only by alkaline solution at comparatively low pressure and temperature was investigated. This novel process is also a cleaner process, where the excess alkaline solution reactant is recycled to the leaching of fly ash, followed by the crystallizing of sodium aluminate hydrate and then its seed decomposing to prepare aluminum hydroxide, calcined to produce alumina. By controlling the phase transformation of the multiphase complex components in the fly ash, overall more than 90% of alumina was recovered from fly ash. The final residues could be utilized as feedstock in construction materials. Furthermore, the reaction mechanism of the process is investigated. This mild and cleaner process could also be appropriate for the produce of alumina from other high Al-containing solid wastes.

(550a) Polymer Reaction Engineering

Choi, K. Y., University of Maryland, College Park

Originally presented on: 11/6/2013 15:20:00 - 15:45:00

Polymer reaction engineering research aims at solving scientific and technical problems relevant to industrial polymerization process technology. Studies of polymerization kinetics, polymerization reactor/process modeling, control, and optimization have been the major research goals. As polymer industry matures, polymer reaction research is now shifting towards the solution of nanoscale polymerization problems for various new applications where phase heterogeneity complicates the process. Our recent investigation includes catalytic heterogeneous reactions to develop silicon-CNT nano-composites as anodes for Lithium ion batteries, micro-channel reactors for heterogeneous polymer reactions, and ultrahigh molecular weight step-growth polymers in confined reaction space. Some challenging polymer reaction problems will be discussed at the meeting.

(550b) Crystallization Technology Using Taylor Vortex Fluid Motion

Kim, W. S., Kyung Hee University

Originally presented on: 11/6/2013 15:45:00 - 16:10:00

The crystallization is a phenomenon that a crystalline solid of solute is formed out of the solute-dissolved solution or gas via nucleation and growth under a supersaturated condition. Thus, the crystallization technology is frequently used for the separation/purification and particle/powder processes in wide areas of industries. In my lab for the last two decades, the principles of the crystallization have been intensively investigated for their applications to the separation of chiral isomers, purification of the bio-products, polymorphs and phase transformation of the active pharmaceutical ingredients, synthesis of nano-particles etc. Among those, new crystallization technology using the Taylor vortex fluid motion has been the hottest subject for the last two decades.

The Taylor vortex fluid motion is induced in the gap between two co-axial cylinders when rotating the inner cylinder. Even though this fluid motion was found over 100 years ago, its application to the crystallization was first attempted in my lab to the gas(CO_2)-liquid($Ca(OH)_2$) reaction crystallization of calcium carbonate twenty year ago. In this study, it was found that the mass transfer rate at the gas-liquid interface in the Taylor vortex fluid motion was 3-4 times faster than that in the random eddy fluid motion, producing the nano-size spindle calcite particles in the Taylor vortex crystallizer. In the anti-solvent crystallization of guanosine mono-phosphate, the Taylor vortex crystallizer in continuous mode demonstrated the 100 times in crystal productivity over conventional crystallizer in the batch mode. In addition, the periodic Taylor vortex aligned the solute molecules in the solution, resulting in the polymorphic nucleation of stable form of the sulfamerazine, which was not possible in the random eddy flow. Further, the high flexibility in operation of the Taylor vortex crystallizer such as distributive feeding strategy and temperature gradient technique was developed for the high crystal productivity and the narrow crystal size distribution. By the study for two decades, the high applicability of the Taylor vortex to the crystallization technology in variety of the material processing was demonstrated in my lab.

(550c) Towards Sustainable Energy: Novel Liquid-like Organic Hybrid Nanomaterials for CO2 Capture and Conversion

Park, A. H. A., Columbia University

Originally presented on: 11/6/2013 16:10:00 - 16:35:00

The rapidly-increasing concentration of Greenhouse Gases, particularly CO_2, in the atmosphere has become one of the most difficult challenges faced by humanity, while the use of carboneous fuels is still important for the health of the global communities. Therefore, various carbon capture and storage technologies are currently being developed to meet this challenge. Amine-based aqueous solvents represent the most mature CO_2 capture method but their large-scale deployment is still compromised because of their highly volatile nature, tendency to degrade, corrosiveness and high energy penalty associated with the solvent regeneration. To address these issues, the development of an alternative CO_2 capture solvent has become a very active research area. Liquid-like Nanoparticle Organic Hybrid Materials (NOHMs) are one of these novel CO_2 solvents with a number of interesting chemical and physical properties. NOHMs are a new class of organic-inorganic hybrids that consist of a hard nanoparticle core functionalized with a molecular organic (e.g., polymeric) corona that possesses a high degree of tunability. NOHMs are non-volatile and stable over a very wide temperature range, which make them interesting materials for various energy and environmental applications. The enthalpic and entropic mechanisms of CO_2 sorption in NOHMs will be discussed along with parameters relevant to CO_2 capture technologies, e.g. selectivity, recyclability, thermal stability or viscosity.

(550d) Panel: Looking Ahead in Developing Academic or Industrial Career

Joo, Y. L., Cornell University
Choi, K. Y., University of Maryland, College Park
Moon, I., Yonsei University
Chun, J., Pacific Northwest National Laboratory
Lee, J. S.,

Originally presented on: 11/6/2013 16:35:00 - 17:10:00

Academic career in the U.S. institutions would be one of the most exciting but challenging careers, especially for foreign-born scholars in science and engineering. Modern university administrators demand young academics for supremacy in research, funding, and teaching regardless of the amount of support they may provide. Academic research community is also rapidly globalizing and more competitive than ever. At the same time, industrial research support or collaboration is rapidly disappearing. In this forum, we will discuss some issues relevant to the sustainability in academic profession in this harsh environment.

(550e) Award Ceremony

Kim, S. H., The Pennsylvania State University

Originally presented on: 11/6/2013 17:10:00 - 17:25:00

Awards:

1) Doh Won Seok Award (administrated by Prof. Dong Sik Kim, Tulane University)

2) Service recognition award: Prof. Yong L. Joo (former KIChE-US chapter president)

3) Sponsor Acknowledgement: SK Innovations

4) KIChE president's award: Hyunmin Yi (Tufts University)

(550f) Nanobiofabrication: Exploiting Programmable Properties of Biological Materials and Interactions for Fabrication of Nanocatalysts and Biosensing Platforms

Yi, H., Tufts University

Originally presented on: 11/6/2013 17:25:00 - 17:45:00

Viral assemblies have attracted substantial attention as templates for materials synthesis due to their precisely controlled dimensions, chemical functionalities and the ability to confer additional modalities through genetic modification. At the nanobiofabrication group of Tufts University, we exploit several unique properties of tobacco mosaic virus (TMV) for facile synthesis of catalytically active palladium (Pd) nanoparticles and fabrication of high capacity biosensing platforms. In the nanocatalysis area, we have examined and demonstrated size-controlled synthesis, high thermal stability and the TMV template's fundamental role in the Pd nanoparticle formation via small angle X-ray scattering (SAXS). We then employed two Pd-catalyzed model reactions, dichromate reduction for environmental cleanup and Suzuki coupling reaction for efficient chemical synthesis, in order to investigate the catalytic activity, stability, and reaction mechanisms. The results show that TMV-templated Pd nanoparticle synthesis offers attractive routes to highly active, controlled, and stable catalyst systems in mild aqueous conditions. In the biosensing area, we have enlisted TMV's dual functionality coupled with robust replica molding (RM) for facile fabrication of high-capacity biomolecular assembly platforms. First, genetically displayed cysteines allow for biomolecular conjugation via copper-free click chemistry. Second, partial disassembly of TMV's mRNA genome enables orientationally controlled assembly via nucleic acid hybridization. Third, simple and robust RM technique leads to scalable fabrication of shape-controlled polymeric microparticles. We envision that our approach to seamlessly integrate these three traits would lead to promising fabrication schemes for high-capacity biosensing platforms. In this presentation, our recent progress on the spontaneous Pd particle formation, integration of Pd-TMV complexes into polymeric scaffolds, and fabrication of hybrid microparticles with chitosan oligomers as an efficient biomolecular assembly modality will also be highlighted.

(823a) Strategic Imperatives for the Hydrocarbon Industries
Gautam, R.,
Originally presented on: 11/6/2013 18:00:00 - 18:25:00

As the leading international supplier of catalysts, adsorbents, equipment and process technology for the petroleum refining, gas processing, petrochemicals and renewable energy industries, UOP has pioneered and implemented some of the most innovative and productive technologies in use today. We are now on the brink of momentous changes in our industry. Feedstocks derived from natural gas, biomass and coal are finding a foothold in the production of fuels and petrochemicals and new processing routes are being proven and gaining commercial acceptance. This is an exciting time for researchers, technology developers and anyone entering the field of chemical engineering, and the next few years will provide many new opportunities and challenges. The AGILE Award Lecture will provide a strategic survey of the major issues facing the hydrocarbon industries and discuss how chemical engineers are applying new science and technologies to address some of these challenges.

(624a) Carbon Nanomaterial Dense Layer Formation On Graphite Particles for Li-Ion Battery Anode By Using Colloidal and CVD Combined Process

Kamiya, H., Tokyo University of Agriculture and Technology
Nomura, Y., Tokyo University of Agriculture and Technology
Iijima, M., Tokyo University of Agriculture and Technology
Anoshkin, I. V.,
Nasibul, A. G.,
Kauppinen, E.,
Originally presented on: 11/7/2013 8:30:00 - 8:53:00

Carbon nanomaterials, such as carbon nanotube, CNT and carbon nano fiber, CNF, have various high functional properties to develop Li-ion battery performance. However, hard aggregates of CNT during CVD preparation process were formed in gas phase, and such aggregates hindered to prepare uniform and high performance electrode. Such hard aggregates were not able to be dispersed by conventional physical processes. In this research, new preparation process of carbon nanomaterials uniform and dense layer on the surface of graphite fine particles, which is Li ion battery anode raw material, are developed. Uniform carbon nanomaterials coated layer on graphite will develop the performance of Li battery anode material. Firstly, surface modified CoO nanoparticles, which are precursor of Co nanoparticle catalyst for CNT formation in CVD process, were prepared in liquid phase and adsorbed on graphite particles. By the selection of optimum surface organic molecular structure on CoO nanoparticles, high coverage of CoO nanoparticles on graphite was achieved in toluene. Surface interaction control between CoO nanoparticles and graphite was useful to increase the number density of adsorbed nanoparticles on graphite. After reduction of CoO nanoparticles on graphite in gas phase, carbon nano fiber dense layers were formed on the surface of graphite particles by CVD process. Since the amount of produced carbon nano fiber on graphite was depended on the number density of adsorbed CoO nanoparticles, colloidal surface interaction design between CoO nanoparticles and graphite was important for high performance raw powder materials.

(624b) Insertion and Immobilization of Pre-Synthesized Particles Into Sub-100 Nm Porous Structures By Electrophoresis

Kusdianto, K., Tokyo University of Agriculture and Technology
Sasaki, K.,
Lenggoro, W.,
Originally presented on: 11/7/2013 8:53:00 - 9:16:00

Pulse and direct current (DC) electrophoretic deposition of nanometer-sized particles (average sizes of 10 and 50 nm) into sub-100 nm scaled pore arrays made from anodized aluminum has been investigated. The deposited particles were observed when several parameters such as pH, applied voltage, and particle size were altered. At the applied voltages lower than the decomposition voltage of water (1 V), the number concentration of the particles deposited by conventional (constant) DC was higher than that of pulse DC. The number of deposited particles increased with increasing pH. The deposition efficiency inside the pores was enhanced by applying pulse DC. When the voltage as high as 10 V was applied however, no particles were observed inside the pores even by application of pulsed DC. The adhesion strength (removal behavior) of the deposition was evaluated by application of a particle-detachment system based on ultrasonic energy on the structure. The particles deposited inside the pores in the 1 V samples were not found to be detached in contrast to those on the surface.

(624c) Development Of a Novel Rotating Fluidized Bed For Fine Particle Processing

Watano, S., Osaka Prefecture University
Nakamura, H., Osaka Prefecture University
Originally presented on: 11/7/2013 9:16:00 - 9:39:00

A rotating fluidized bed (RFB) which imparts high centrifugal force to fluidizing particles by horizontally rotating the vessel is a promising technology for fine particles processing. We have already reported that the RFB can be used for uniform fluidization of nano particles, micro granulation/ fine particle coating, synthesis of nano particles and etc. In this study, new devices such as a tapered rotating vessel and an inclined injection of fluidizing air have been developed for the better performance of the RFB. The experimental results confirmed that the tapered rotating vessel improved the particle recovery and the inclined injection greatly improved the particle mixing performance. A numerical analysis using a DEM (Discrete Element Method) and CFD (Computational Fluid Dynamics) was also conducted to analyzed the mechanism how the tapered rotating vessel and the inclined injection improved the particle behaviors and performance. The optimum geometry of the devices was also determined by using the numerical analysis.

(624d) Numerical and Experimental Study of the Correlation Between Bulk Stresses and Interparticle Force in the Biomass Combustion Ash

Kamiya, H., Tokyo University of Agriculture and Technology
Tong, Z.,
Originally presented on: 11/7/2013 9:39:00 - 10:02:00

In recent years there has been high commercial interest in the energy production based on combustion of biomass according to future fuel dependency and in order to keep CO_2-emissions down. However, uncontrolled ash deposit build up on the furnace wall which is necessary to be cleaned at regular intervals lead to long periods without power production and large cleaning costs. The consequent lack of basic understanding of the deposition mechanism is a major hindrance to advance of biomass combustion technology. This paper presents measurements of tensile strength as a function of temperature (25-800º) for ash deposit of palm residues combusted at 750 and 820º. It has been shown that the tensile strength is strongly dependent on the temperature. It is interesting that the ash deposit combusted at 750º slightly increase with temperature while the ash deposit combusted at 820º first sharply increase and then sharply decrease with temperature. In order to fundamentally understand the relationship of macroscopic strength and microscopic adhesion, the Discrete Element Method (DEM) simulations were performed with different primary particle size distribution, bulk solid fraction and bond type (related to temperature). The solid bond and liquid bridge between the particles are modeled using the Bonded Particle Model (BPM) and Capillary Force Model (CFM), respectively. The quantitative information obtained from the numerical simulation can be used to better understand the biomass ash deposition mechanisms.

(624e) Hydrodynamic Behaviors of Sand and Plastic Particles in a Large-Scale Triple Bed Combined Circulating Fluidized Bed

Fushimi, C., Tokyo University of Agriculture and Technology
Ishizuka, M., The University of Tokyo
Guan, G., Hirosaki University
Suzuki, Y., National Institute of Advanced Industrial Science and Technology
Norinaga, K., Kyushu University
Hayashi, J. I., Kyushu University
Tsutsumi, A., The University of Tokyo
Originally presented on: 11/7/2013 10:02:00 - 10:25:00

Hydrodynamic behaviors of binary mixture of silica sand and nylonshot (coal char substitute) were investigated in a triple-bed circulating fluidized bed (TBCFB). The TBCFB consisted of a downer (Φ 0.1 m × 6.5 m), a bubbling fluidized bed (0.75 m × 0.27 m × 1.9 m), a riser (Φ 0.1 m × 16.6 m) and a gas-sealing bed (GSB, Φ 0.158 m × 5 m). The experiments were carried out at ambient temperature and atmospheric pressure by using air. The initial fraction of the nylonshot in the solid mixture ($X_{nylon,i}$) was 15.4% and 30% on mass and volume bases, respectively, or otherwise, 30.7% and 50%. The maximum obtained solids mass flux (G_s) at $X_{nylon,i}$ of 15.4 and 30.7wt% were 394 and 349 kg/m^2 s, respectively, when the gas velocity in the riser (U_{gr}) was 10 m/s. Apparent solids holdups of silica sand and nylonshot were calculated separately from the static pressure gradient across the riser and the downer. The results showed that it is possible to circulate char in the gasifier, which plays a significant role in decomposition of tar from pyrolysis as the primary step of gasification. A novel developed pressure balance model successfully predicted Gs of the binary mixtures in TBCFB well.

(624f) Microscopic Numerical Investigation of Dense Gas-Solid Flows Near a Solid Wall (Effects of wall on the behavior of particles)

Fujihara, S., Osaka University
Tsuji, T., Osaka University
Tanaka, T., Osaka University
Originally presented on: 11/7/2013 10:25:00 - 10:48:00

Flow including dense solid particles often appears in industrial equipments such as fluidized beds and spouted beds used for combustion, catalytic reaction, granulation and coating processes. To enable advanced designing and optimization of operating conditions, it is important to understand the flow behavior in the equipment in detail. However, its behavior is complicated especially in the near-wall region. A solid wall affects the flow directly by wall friction effect and the particle arrangement through the contacts between wall-particle and particle-particle. Due to the difficulty of experimental observations, the effect of wall is not known well at present. In the present study, the influence of solid wall on a flow field including dense moving particle is investigated in detail by means of microscopic direct numerical simulation. Immersed boundary method is combined with discrete element method. In order to investigate the influence of solid wall, sedimentation of dense particles near a vertical wall is simulated. To perform the computations with reduced computational cost, a new mirror boundary condition is proposed, too. It is confirmed that in near-wall region, the existence of the solid wall has significant effects both on particle and fluid flow structures. Moreover, sedimentation velocity of particles near solid wall is decreased and fluid resistance is increased due to the wall.

(624g) Drag Force Relation By Direct Numerical Simulation for Gas-Solid Flows

Zaidi, A., OSAKA UNIVERSITY
Tsuji, T., Osaka University
Tanaka, T., Osaka University
Originally presented on: 11/7/2013 10:48:00 - 11:11:00

Direct Numerical simulations are used to propose the drag correlation for random mono-disperse spheres which are applicable to high stokes number particles which can be used for the estimation of gas-solid flows. The simulated solid volume fraction varies from 0.05 to 0.5 and Reynolds number varies from 0.01 to 1000. To make sure that the current results are reliable, care is taken in the selection of parameters of simulation (e.g. grid resolution, domain size etc.). Average drag force acting on particles in both regular and random arrangements is compared with literature data and good agreement is observed. One particular exception was the simulation results of Beetstra et al. (2007) who used Lattice Boltzmann method which showed large deviation with the current simulations. We think that the grid used by Beetstra is not resolved enough for the range of parameters he studied. At the end we curve fitted our simulation results and proposed a new drag relation. We expect that this new relations can help to improve the numerical modeling of gas–solid flows in mesoscopic simulations.

(648a) Background on NuClean Initiative

Beloff, B., Beth Beloff & Associates
Calabrese, R. V., University of Maryland
Manousiouthakis, V., University of California Los Angeles, Los Angeles
Arm, S. T., One System
Originally presented on: 11/7/2013 9:00:00 - 9:45:00

Goals: Establish *NuClean* as a nuclear waste management center of excellence to leverage chemical engineering expertise on nuclear waste management and engage with broader professional efforts on underlying multi-disciplinary public and technical community information, research, outreach and education needs.

NuClean ***will*** be an unbiased forum for multiple stakeholders to develop a common understanding of concerns and expectations associated with nuclear waste management, from both commercial and defense activities, legacy and present, and to encourage education and professional development of stakeholders with respect to these topics.

NuClean ***will*** be a source for best practices in risk assessment, education and development and application of technologies for protecting human health and the environment related to nuclear waste management and disposal, as well as potential 'sustainable' nuclear fuel cycle technologies. *NuClean* potentially can host applicable communities of practice.

Approach: To accomplish this *NuClean* must be collaborative, cross-disciplinary and cross-sectoral, with experts from technology, public health, social sciences, business, academic, government, community and NGOs working together. Current efforts to address community concerns are fragmented, so there is a need for an inclusive and integrated effort or systems approach to bring together the best solutions to be transferred widely.

Rationale: Globally, concerns over safety of nuclear waste are driving peoples' perception of the risks posed by nuclear power. There is a need for a neutral professional, public voice that can provide trusted, factual information to diverse stakeholders and communities regarding safety issues and uncertainties associated with management of nuclear waste from both defense legacy and civilian energy to improve understanding of the issues, drive development of solutions, and increase confidence in nuclear waste processing, storage and handling. Furthermore, there have been clearly identified shortfalls in current and projected future availability of professionals educated in critical areas supporting national needs in nuclear materials processing and waste management that align well with core chemical engineering expertise, including process safety, process development, analysis and simulation, chemical separations, waste treatment and environmental remediation. These two needs, sources of reliable factual information accessible to the public, and, workforce/education provide the basis for greater emphasis and initiatives by AIChE on these topics.

Background: By way of an example, many communities downstream of Los Alamos National Laboratory have concerns about the migration of toxic and radioactive materials as a result of long-term nuclear waste storage. With increasing wildfires around LANL and subsequent erosion from flooding, there are growing concerns regarding the migration of legacy waste into drinking water, onto agricultural lands, and into the tissue of humans and other members of affected ecosystems. Not only is quality of life affected by these concerns, but also economic development may be impacted.

New Mexico stakeholders, however, are not unique in their concerns about the risks posed by nuclear waste management. Similar perceptions exist at Hanford, WA, Savannah River, NC and in Idaho, as well as with respect to decommissioning of civilian power reactors. As the nation responds to a growing energy demand in a climate neutral way, a population exists that must be informed of the safety and environmental risks of nuclear waste and how they can be mitigated. However, current issues about the safety and environmental security of nuclear waste need to be addressed responsibly today for future generations.

(648b) Overview: Nuclear waste management - Technical, Political, Regulatory Challenges regarding Commercial and Legacy Waste

Gilbertson, M., Department of Energy
Kosson, D., Vanderbilt University
Originally presented on: 11/7/2013 9:45:00 - 10:30:00

Overview: Nuclear waste management - Technical, Political, Regulatory Challenges regarding Commercial and Legacy Waste, including the state-of-the-art in technologies and organizations for responding to current nuclear waste processing, storage and remediation issues and the US Science Policy

Mark Gilbertson, DOE Deputy Assistant Secretary, Site Restoration for Environmental (invited);David Kosson, Vanderbilt University, Cornelius Vanderbilt Professor of Engineering, Director of Consortium for Risk Evaluation with Stakeholder Participation (CRESP)

(648c) Assessment and Perception of Risks Associated with Nuclear Waste

Jones, E., Lawrence Livermore National Laboratory
Greenberg, M., Rutgers University
Whipple, C. G.,
Naranjo, M.,
Originally presented on: 11/7/2013 11:00:00 - 11:45:00

Assessment and Perception of Risks Associated with Nuclear Waste, including how the public perceives these risks and need for communities to know about waste streams and their consequences

Ed Jones, LLNL, Deputy Associate Program Leader for Risk Assessment, System Engineering, and Human Performance; Michael Greenberg, Rutgers, Faculty and Associate Dean, Edward J Bloustein School of Planning and Public Policy, Director National Center for Neighborhood and Brownfields Redevelopment (invited); Marian Naranjo, HOPE (Honor Our Pueblo Existence) (invited)

(649a) Clicking Polymers Together: Assembly of Complex, Controlled Polymer Structures from Efficient Chemistries

Bowman, C. N., University of Colorado

Originally presented on: 11/7/2013 11:15:00 - 11:40:00

A new paradigm encompassing several distinct chemical reactions and, more importantly, a generalized approach to molecular design and synthesis has been rapidly adopted in the fields of chemical synthesis, biotechnology, materials science, drug discovery, surface science, and polymer synthesis and modification. The *Click Chemistry* paradigm focuses on implementation of highly efficient reactions that achieve quantitative conversion under mild conditions. As such, these reactions represent ideal candidates for further development, understanding and implementation. In particular, the synergistic combination of these click chemistries with photochemical initiation and polymer formation has been used to afford 4D control of polymer formation, structure and patterned assembly. Here, we will focus on three distinct vignettes related to our implementation of photoclickable polymer systems. The first of these focuses on the development of covalent adaptable networks (CANs) where the ability to controllably alter the network structure is used to alter topography and other material properties by forming materials which can be switched revesibly from elastic to plastic simply by exposure to light. Secondly, we will focus on the development of approaches to photoinitiate the Cu(I) catalyzed azide-alkyne cycloaddition (CuAAC) click reaction. Here, implementation of this reaction in surface modification, hydrogel formation, and lithography as well as in the development of a new class of photopolymerization reactions will be presented. Finally, the development and implementation of click nucleic acids (CNAs) based on the thiol-ene click reaction will be presented. This distinct class of oligonucelotides combines the vast advantages of synthetic oligonucleotides such as peptide nucleic acids with the power of click reaction chemistry to form materials that hybridize with both natural and synthetic olignonucleoitides via Watson-Crick base pairing while being simple to produce in large scales appropriate for directed assembly and other high value materials applications.

AUTHOR INDEX

AUTHOR INDEX

AUTHOR INDEX